Lecture Notes in Bioinformatics   13063

Subseries of Lecture Notes in Computer Science

More information about this subseries at http://www.springer.com/series/5381

Peter F. Stadler · Maria Emilia M. T. Walter ·
Maribel Hernandez-Rosales ·
Marcelo M. Brigido (Eds.)

Advances in Bioinformatics and Computational Biology

14th Brazilian Symposium on Bioinformatics, BSB 2021
Virtual Event, November 22–26, 2021
Proceedings

Springer

Editors
Peter F. Stadler ⓘD
Leipzig University
Leipzig, Sachsen, Germany

Maria Emilia M. T. Walter ⓘD
University of Brasília
Brasilia, Brazil

Maribel Hernandez-Rosales ⓘD
CINVESTAV
Irapuato, Mexico

Marcelo M. Brigido ⓘD
Universidade de Brasilia
Brasilia, Brazil

ISSN 0302-9743   ISSN 1611-3349 (electronic)
Lecture Notes in Bioinformatics
ISBN 978-3-030-91813-2   ISBN 978-3-030-91814-9 (eBook)
https://doi.org/10.1007/978-3-030-91814-9

LNCS Sublibrary: SL8 – Bioinformatics

This Springer imprint is published by the registered company Springer Nature Switzerland AG
The registered company address is: Gewerbestrasse 11, 6330 Cham, Switzerland

Preface

The Brazilian Symposium on Bioinformatics (BSB) is an international conference that covers all aspects of bioinformatics and computational biology. This volume contains the accepted papers for BSB 2021, held virtually during November 22–26, 2021.

As in past years, the special interest group in Computational Biology (CEBioComp) of the Brazilian Computer Society (SBC) organized the event. A Program Committee (PC) was in charge of reviewing submitted papers; this year, the PC had 42 members. Each submission was reviewed by three PC members. There were two submission tracks: full papers and short papers. In the full paper track, ten papers were accepted; five papers were accepted in the short paper track. All of them are printed in this volume and were presented orally at the event. In addition to the technical presentations, BSB 2021 featured the following invited speakers, with the respective talk titles: João Carlos Setubal (University of São Paulo, Brazil), Microbial genome informatics in the microbiome era; Angelica Cibrian (CINVESTAV, Mexico), Evolutionary genomics of ancient plants and their symbionts: from species to metabolites; Marc Hellmuth (Stockholm University, Sweden), Gene family histories and homology relations; and Deisy Morselli Gysi (Northeastern University, USA), Drug repurposing for treating diseases: a network medicine approach. Also, we organized a special session dedicated to "Bioinformatics and Artificial Intelligence" coordinated by Maribel Hernandez-Rosales (CINVESTAV Irapuato, Mexico). Moreover, we proposed three round tables with the objective of exchanging ideas among students, researchers, and professionals, which were coordinated by following researchers focusing on the specific themes: Peter Stadler (Leipzig University, Germany), Quo vadis, bioinformatics/computational biology?; Maribel Hernadez-Rosales (CINVESTAV Irapuato, Mexico) and Vinita Gowda (IISER Bhopal, India), STEM questions and challenges in bioinformatics; Steve Hoffman (Leibniz Institute on Aging Research/Fritz Lipmann Institute, Germany), Elizabeth Tapia (CIFASIS-CONICET-UNR, Argentina), and Tania Carrillo-Roa (Roche Diagnostics), Career opportunities - academy and industry.

BSB 2021 was made possible by the dedication and work of many people and organizations. We would like to express our thanks to all Program Committee members. Their names are listed in the pages that follow. We are also grateful to the local organizers, Raquel Minardi, Waldeyr M. C. Silva, Daniel de Oliveira, Marcelo Reis, and volunteers for their valuable work and for helping out with outreach; to the sponsors for making the event financially viable; to the maintainers of JEMS, which was the system we used to handle submissions; and to Springer for agreeing to publish this volume and their staff for working with us on its production. Last but not least, we would like to thank all authors for their time and effort in submitting their work and the invited speakers for having accepted our invitation.

November 2021

Peter F. Stadler
Maria Emilia M. T. Walter
Maribel Hernandez-Rosales
Marcelo M. Brigido

Organization

General Chair

Raquel Minardi · Federal University of Minas Gerais, Brazil

Organization Committee

Daniel Cardoso Moraes de Oliveira · Fluminense Federal University, Brazil
Waldeyr Mendes Cordeiro da Silva · Federal Institute of Goiás, Brazil
Marcelo da Silva Reis · Butantan Institute, Brazil

Program Committee Chairs

Peter F. Stadler · Leipzig University, Germany
Maria Emilia M. T. Walter · University of Brasília, Brazil
Maribel Hernandez-Rosales · CINVESTAV Irapuato, Mexico
Marcelo M. Brigido · University of Brasília, Brazil

Steering Committee

Daniel Cardoso Moraes de Oliveira · Fluminense Federal University, Brazil
Waldeyr Mendes Cordeiro da Silva · Federal Institute of Goiás, Brazil
João Carlos Setubal · University of São Paulo, Brazil
Luis Antonio Kowada · Fluminense Federal University, Brazil
Raquel Minardi · Federal University of Minas Gerais, Brazil
Ronnie Alves · Vale Institute of Technology, Brazil
Sérgio Campos · Federal University of Minas Gerais, Brazil
Sérgio Lifschitz · Pontifical Catholic University of Rio de Janeiro, Brazil

Program Committee

Adriano Werhli · Federal Universidade of Rio Grande, Brazil
Alexandre Paschoal · Federal University of Technology - Paraná, Brazil
André Ponce de Leon F. de Carvalho · University of São Paulo, São Carlos, Brazil
Andre Kashiwabara · Federal University of Technology - Paraná, Brazil
César Benítez · Federal University of Technology - Paraná, Brazil

Christian Hoener zu Siederdissen	University of Jena, Germany
Clara Bermudez Santana	National University of Colombia, Colombia
Danilo Sanches	Federal University of Technology - Paraná, Brazil
Fabrício Lopes	Federal University of Technology - Paraná, Brazil
Fábio Custódio	National Laboratory of Scientific Computation, Brazil
Felipe Louza	Federal University of Uberlândia, Brazil
Glauber Wagner	Federal University of Santa Catarina, Brazil
Guilherme Telles	University of Campinas, Brazil
Guillermo Jáuregui	Inmegen, Brazil
Gustavo Teixeira Chaves	Federal Institute of Goiás, Brazil
Jefferson Morais	Federal University of Pará, Brazil
Joao Carlos Setubal	University of São Paulo, Brazil
Karina dos Santos Machado	Federal University of Rio Grande, Brazil
Kary Ocaña	National Laboratory for Scientific Computing, Brazil
Luciano Digiampietri	University of São Paulo, Brazil
Luis Cunha	Fluminense Federal University, Brazil
Luiz Manoel R. Gadelha Júnior	National Laboratory for Scientific Computing, Brazil
Marcelo M. Brigido	University of Brasília, Brazil
Marcilio de Souto	Université d'Orléans, France
Marcio Dorn	Federal University of Rio Grande do Sul, Brazil
Maria Emilia M. T. Walter	University of Brasília, Brazil
Mariana Recamonde-Mendoza	Federal University of Rio Grande do Sul, Brazil
Maristela Holanda	University of Brasília, Brazil
Nalvo F. Almeida Jr.	Federal University of Mato Grosso do Sul, Brazil
Raquel Costa	Brazilian National Cancer Institute, Brazil
Rommel Thiago Ramos	Federal University of Pará, Brazil
Ronnie Alves	Vale Institute of Technology, Brazil
Sabrina Silveira	Federal University of Viçosa, Brazil
Said Adi	Federal University of Mato Grosso do Sul, Brazil
Sergio Campos	Federal University of Minas Gerais, Brazil
Sergio Lifschitz	Pontifical Catholic University of Rio de Janeiro, Brazil
Steve Hoffmann	Leibniz Institute on Aging/Fritz Lipmann Institute, Germany
Thaís Gaudêncio do Rêgo	Federal University of Paraíba, Brazil
Thais Amaral e Sousa	Federal Institute of Goiás, Brazil
Valerie Anda	University of Texas, USA
Waldeyr M. C. Silva	Federal Institute of Goiás, Brazil
Zanoni Dias	University of Campinas, Brazil

Financial Support

National Council for Scientific and Technological Development, Brazil
Pontifical Catholic University of Rio de Janeiro, Brazil
RIABIO Ibero-American Network on Artificial Intelligence Applied to Big BioData

Sponsor

Brazilian Computer Society (Sociedade Brasileira de Computação), Brazil

Contents

Applied Bioinformatics
and Computational Biology

Comparative Transcriptome Profiling of *Maytenus ilicifolia* Root and Leaf

Mariana Marchi Santoni[1]([✉])(iD), João Vítor Félix de Lima[1](iD),
Keylla Utherdyany Bicalho[2](iD), Tatiana Maria de Souza Moreira[1](iD),
Sandro Roberto Valentini[1](iD), Maysa Furlan[3](iD), and Cleslei Fernando Zanelli[2](iD)

[1] Department of Biological Sciences, School of Pharmaceutical Sciences, São Paulo
State University (UNESP), Araraquara, SP 14800-903, Brazil
mariana.santoni@unesp.br
[2] VIB Center for Plant Systems Biology, 9052 Ghent, Belgium
[3] Department of Organic Chemistry, Institute of Chemistry, São Paulo State
University (UNESP), Araraquara, SP 14800-900, Brazil

Abstract. Plants produce a wide variety of compounds called secondary
metabolites (SMs), which are extremely important for their survival. SMs
have also medicinal applications, but as chemical synthesis is not economically viable, plant extraction is the mainly option. Different biotechnology strategies are applied to improve the yield of bioproduction of these
compounds, but commonly without the desired results due the limited
knowledge of biosynthetic and regulatory pathways. *Maytenus ilicifolia*,
a traditional Brazilian medicinal plant from Celastraceae family, produces in both root and leaves three main classes of SMs: sesquiterpenics,
flavonoids and quinonemethides. In this study, four cDNA libraries were
prepared from root and leaf tissues. The *de novo* transcriptome included
109,982 sequences that capture 92% of BUSCO orthologs, presented an
average length of 737bp and a GC content about 42% of. Function annotation analysis identified homology for 44.8% of the transcripts. Moreover, 67,625 sequences were commonly expressed in both tissues, while
1,044 and 1,171 were differentially expressed in root and leaf, respectively. In terms of SM, enzymes involved in "monoterpenoid biosynthesis"
and "isoflavonoid biosynthesis" were identified in root while "flavonoid
biosynthesis" and "Biosynthesis of alkaloids" in leaf.

Keywords: RNA-Seq · *De novo* assembly · Metabolic pathways

1 Introduction

Compounds produced by plants are categorized into primary and secondary
metabolites (SMs). Primary metabolites, such as carbohydrates, lipids, and proteins, are involved in plant development [9] and essential for cell growth [17]. In

Supported by São Paulo Research Foundation (FAPESP) [2013/07600-3; 2016/16970-
7]; National Council for Scientific and Technological Development (CNPq)
[303757/2017-5]; National Institute for Science and Technology (INCT).

P. F. Stadler et al. (Eds.): BSB 2021, LNBI 13063, pp. 3–14, 2021.
https://doi.org/10.1007/978-3-030-91814-9_1

contrast, SMs (low molecular weight compounds) are multifunctional metabolites produced as an evolutionary adaptation [4,23]. They are involved in plant defense and environmental communication [4,8,23], plant color, taste, and scent [9] and responses to biotic and abiotic stress [15,16,19].

The high variety of biological functions of the SMs is explained by diversified chemical structure [4,24] originated from a restricted and distinct number of metabolic pathways such as the acetate, shikimic acid, mevalonic acid or methylerythritol phosphate pathways [9,21]. SM are grouped in three classes: terpenes, alkaloids and phenylpropanoids, each one with its respective and unique properties [4,24]. These compounds are identified in all plant tissues and their formation and gene regulation is usually organ, tissue, cell and also development specific, indicating that a range of transcription factors must cooperate to transcribe secondary metabolism genes, controlling the general machinery of biosynthetic pathways in production, transport and storage [18,22].

Many SMs are sources of drugs however, as chemical synthesis is uneconomical, isolation from plants still represents the only option [4,13]. Different biotechnological strategies have been applied to improve the production of these compounds, but often without the desired results due to the lack of knowledge about the biosynthetic routes [13,21]. Biotechnology techniques such as transcriptome, proteome or metabolomics are used to identify genes and their functions in plant metabolic pathways in order to clarify the mechanisms involved in SMs synthesis [4].

Maytenus ilicifolia Mart ex Reissek (Celastraceae) is a Brazilian native plant known for its variety of therapeutic properties. It has been used as a treatment of several diseases such as gastric ulcer, dyspepsia, stomach acidity, diabetes and cancer [12,13,20]. This species produces three main classes of bioactive compounds: alkaloids sesquiterpene pyridines, flavonoids and quinonemethide triterpenes [13] and the mainly products are maitenin, friedelin, fridelanol, pristimerine and terpenes [14]. Additionally, like other members of Celastraceae family, some compounds are synthetized in a specific tissue: quinone methide triterpenoids are accumulated in root bark [1,13] and flavonoids in leaves [2].

The analysis of differentially expressed transcripts between two tissues can provide a better understanding of genes involved in secondary metabolic pathways [3,10,11]. In this context, the aim of the present study was to analyze whole transcriptome of *M. ilicifolia* and identify genes involved in biosynthesis of SMs by a comparative profiling of root and leaf. This study is the first report of high-throughput analysis (*de novo* RNA-Seq) of *M. ilicifolia transcriptome* that provides new insights at molecular knowledge.

2   Methods

2.1   Plant Material and Total RNA Isolation

Leaves of adult specimen of *M. ilicifolia* from the medicinal plant garden of the Faculty of Pharmaceutical Sciences and leaves and roots of identified seedlings, with approximately 6 months of planting, were harvested and stored in −80 °C

(Fig. 1A). The total RNA from two specimens of roots (from two seedlings) and two specimens of leaves (one leaf from seedling, coinciding with one of the specimens used for root extraction and one leaf from adult specimen) was isolated from 500 mg of material using RNeasy Plant mini kit (Qiagen, USA) according to the manufacturer's protocol. RNA quantity and quality were evaluated using Nanodrop 1000 spectrophotometer and Agilent 2100 Bioanalyzer. RNA samples with quality ratios greater than 1.8 (260/280 nm and 28S/18S) and RNA integrity number (RIN) greater than 7 were selected for subsequent processes.

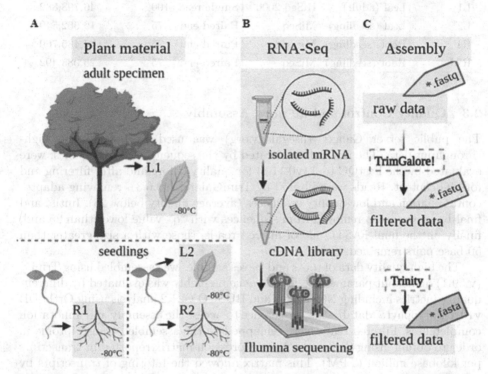

Fig. 1. Experimental approaches for *Maytenus ilicifolia* transcriptome study. A. Two samples of leaves (one leaf from adult specimen and one leaf from seedling) - L1 and L2 - and two samples of roots from two independent seedlings (one root sample coinciding with the same specimen of the leaf sample) - R1 and R2 - were collected and stored in −80 °C for posterior RNA isolation. **B.** Library preparation and transcriptome sequencing **C.** Pipeline used for *de novo* assembly.

2.2 Library Preparation and Sequencing

After isolated from total RNA with magnetic Oligo (dT) particles, mRNA was chemically fragmented. Subsequently, cDNA libraries were prepared using Illumina TruSeq RNA sample preparation v3 kit (Illumina, USA) (Fig. 1B). Quantification and quality assessment of resulting libraries were performed on Agilent 2100 Bioanalyzer. A total of 20 pmol of the libraries was submitted to "single-read" sequencing in HiSeq 2000 platform (leaf of the adult specimen) - FCAV/Unesp - to generate 100bp reads or sequencing in MiSeq equipment (leaf and roots of seedlings) - LAB Multi-FCFAR/Unesp - to generate 75bp paired-end reads (Table 1).

Table 1. RNA-Seq traits of four *Maytenus ilicifolia* (*same specimen).

Sample ID	Plant material	Seq platform	Protocol	Read length	No. of reads
L1	Leaf (adult)	HiSeq 2000	Single read	100	46,798,882
L2*	Leaf (seedling)	MiSeq	Paired-end	75	19,062,549
R1	Root (seedling)	MiSeq	Paired-end	75	24,435,760
R2*	Root (seedling)	MiSeq	Paired-end	75	25,083,492

2.3   Quality Control and *de novo* Assembly

The public server Galaxy (usegalaxy.org) was used to process the high-throughput data. The raw data generated by the sequencing, FASTQ files, were evaluated by the FastQC tool (v0.11.8) for quality before and after filtering and for GC content. Reads were filtered by TrimGalore! (v0.6.3), removing adapter contamination and low-quality sequences (average quality below 25). Initial and final bases were also removed from sequences with "q" value lower than 25 and, finally, in the final FASTQ file of filtered reads, those with a size greater than 50 base pairs remained.

The high-quality data of roots and leaves samples was assembled using Trinity (v2.9.1) on default parameters. The *de novo* assembly was evaluated by different quality metrics including N50 length and BUSCO v4.1.2 analysis using OrthoDB v10 'embryophyta' database as a reference to access the assembly and annotation completeness. Filtered reads were remapped to the assembled transcriptome in order to obtain, using Salmon tool, an expression matrix reported in transcripts per kilobase million (TPM). This matrix allowed the filtering of transcripts by low expression, considering only those with at minimum 1% of dominant isoform expression, generating the filtered transcriptome.

2.4   Functional Annotation

TransDecoder tool was used to find the probable coding regions of transcripts and the open reading frames (ORFs) with a minimum length of 100 amino acids. Then, functional annotation of the transcripts was performed using BLASTX against Uni-ProtKB/SwissProt databases and uniprot _trEMBL _plants database (E-value <1e−5). Moreover, a homology search based on the BLASTP was performed using the predicted proteins as query against UniProtKB/SwissProt databases (E-value <1e−5). The assignments of Gene Ontology (GO) terms to transcripts were performed based on UniProtKB/SwissProt database to assign unigenes to functional categories. Additionally, the proteins with Enzyme Commission (EC) numbers were mapped onto the Kyoto Encyclopedia of Genes and Genomes (KEGG) Pathway Database using online KEGG Automatic Annotation Server (www.genome.jp/kegg/kaas) to assign pathway information to the transcripts.

2.5　Differential Expression Analysis

Salmon tool was applied to estimate the expression level of transcripts. Each filtered FASTQ file was separately aligned to the filtered transcriptome. Then, the expression level of each transcript was normalized and reported in TPM. To summarize the results and provide statistical tests for tissue comparison, the differential expression analysis was performed using DESeq2 R package and transcript expression difference was considered significant when the adjusted p-value < 0.05.

2.6　Gene Ontology Enrichment and KEGG Analysis

Gene ontology (GO) enrichment analysis for biological process (BP) and molecular function (MF) for the differentially expressed transcripts in each tissue was conducted using topGO R package. Significant GO terms (Fisher's exact test p-value < 0.01) were visualized using REViGO (revigo.irb.hr) for semantic space reduction. Transcripts associated with Enzyme Commission (EC) numbers were mapped onto the KEGG pathway database.

3　Results and Discussion

3.1　*De novo* Assembly and Functional Annotation of *M. ilicifolia*

The single-read leaf cDNA library and the paired-end leaf and root cDNA libraries subjected to full transcriptome sequencing generated about 115 million of raw reads. The detailed information of the read numbers in different samples is provided in Table 1.

High quality sequencing data, 112,609,211 reads, was used for assembly. The de novo transcriptome generated included 163,780 transcripts (isoforms) with a GC content of 41.8% and the N50 resulting in 1,222bp. The average transcript size was 737 and 22% of them presented more than 1,000bp (Fig. 2A). By considering transcript expression, 15,704 transcripts represented 90% of the total expression data (Ex90) and had an N50 of 1,487bp (Ex90N50). In addition, the assembled transcriptome of *M. ilicifolia* captured 92.6% of the 1,614 orthologs described for the Virdiplantae database (updated 2020-09-10): 53.0%, 39.6%, 4.2%, and 3.2% of the BUSCO genes were respectively classified as complete single copy, complete duplicate, fragmented and absent. After filtering by low expression, the final transcriptome included 109,982 sequences. These results indicate that the integrity of assembly was high, and the sequencing quality had met the requirements of further analysis.

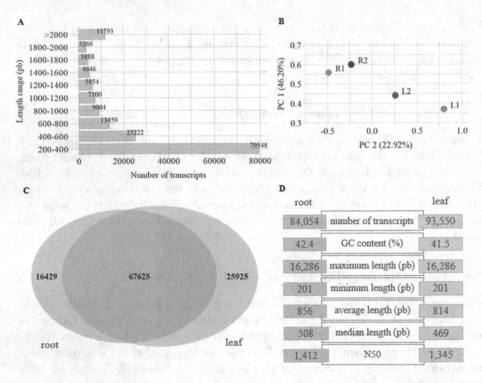

Fig. 2. Aspects of *Maytenus ilicifolia* **transcriptome assembly. A.** Size distribution of assembled data. **B.** Principal component analysis (PCA) on the read counts of root and leaf samples. (L1, L2, R1 and R2 - sample identification described in Fig. 1) **C.** Venn diagram showing the number of transcripts for each source of sample **D.** transcriptome traits for each tissue.

Results of PCA analysis revealed the distinct differences in transcript expression patterns among the samples. The first two principal components contain 69.12% of the information grouping different tissues in separate clusters (Fig. 2B). Considering transcripts identified in leaf and root individually, 67,625 isoforms were found in both tissues (Fig. 2C) and showed similar aspects in respect to transcriptome traits (Fig. 2D).

In summary, M. ilicifolia transcriptome had GC content close to 40%, similar values to those reported for Celastraceae family species like staff vine (41.5%) [20] and thunder god vine (37.2%) [22]. Moreover, results of BUSCO analysis captured more than 90% of the orthologs described for the chosen database and the PCA results allowed the confirmation of expression differences in both tissues, root and leaf.

The BLASTX against the uniprot_trEMBL _plants database found 36,625 alignments and revealed that *M. ilicifolia* predicted transcripts have highest similarity with an organism classified in the same family, Tripterygium sp (47.3%) (Fig. 3A), but homology was find for other family organisms (Fig. 3B). Candidate coding regions in *M. ilicifolia* transcriptome were identified by TransDecoder and 65,533 ORFs and 46,282 probable coding sequences were predicted. Sequence homology search results against the UniprotKB/SwissProt database by BLASTX (E-value <1e−5, for filtered transcripts) and BLASTP (E-value <1e−5, for predicted protein sequences) were 49,319 (44.8%) and 36,344 (55.5%) aligned transcripts, respectively.

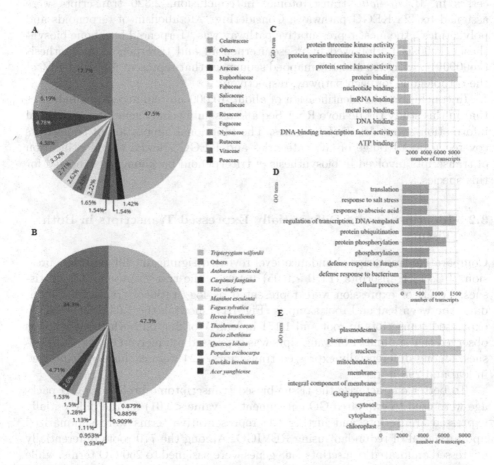

Fig. 3. Functional annotation for *Maytenus ilicifolia* transcriptome. Similarity frequency distribution of different **A.** families and **B.** species. The BLASTx was performed against the trEMBL plants database. Top ten GO terms in the transcriptome assembly from **C.** molecular function, **D.** biological process and

Functional annotation for filtered transcriptome was followed by GO analysis and 43,322 annotated transcripts were categorized into 9,989 GO IDs. The number of transcripts in three main categories of molecular function (MF), biological process (BP) and cellular component (CC) was 41,148, 39,404 and 39,582, respectively. The most dominant GO terms in the MF category were "protein binding,"

"ATP binding" and "metal iron binding" (Fig. 3C). In the BP category, "regulation of transcription", "protein phosphorylation" and "protein ubiquitination" were the most prominent (Fig. 3D). In the CC category, "nucleus", "plasma membrane" and "integral component of membrane" were the most abundant terms (Fig. 3E).

KEGG annotation analysis was performed to identify active metabolic processes in *M. ilicifolia* transcriptome. In conclusion, 2,326 transcripts were assigned to 428 KEGG pathways. Considering "Metabolism of terpenoids and polyketides", the most representative pathway was "Terpenoid backbone biosynthesis (ko00900)", followed by "Sesquiterpenoid and triterpenoid biosynthesis (ko00909)", with 216 and 127 mapped sequences that represent 50% and 10% of the orthologous for each pathway, respectively.

In conclusion, the identification of about 40,000 protein accessions indicates that in this study the de novo RNA-Seq and assembly could generate substantial information about *M. ilicifolia* genes. The functional annotation of transcripts covered a broad range of GO categories and KEGG allowed the identification of transcripts involved in biosynthesis of triterpenoid backbone, as expected for this species.

3.2   Identification of Differentially Expressed Transcripts in Both Tissues

Comparative transcript abundance level revealed significant differential expression of 2,215 transcripts (FDR <0.05) between the transcriptome of both tissues. Levels of expression were represented as $\log_2$ ratio of transcripts abundance between leaf and root samples (Fig. 4A), showing the 1,044 differentially expressed transcripts in root and 1,171 in leaf. Working on both tissues, it was observed that a number of transcripts was expressed uniquely in either of the tissues: among differentially expressed transcripts, 424 were exclusively expressed in leaf and 298 in root.

To better characterize the tissue-biased transcriptome profile, topGO package were used to evaluated GO enrichment (p-value <0.01) for the differentially expressed transcripts and further the representative terms were summarized upon removal of redundant using REVIGO. Among the 770 roots differentially expressed annotated transcripts, 568 genes were assigned to 260 GO terms, while in the leaf, from the 902 differentially expressed annotated transcripts, 610 were classified in 265 GO terms.

The GO analysis revealed enrichment for biological processes (BP) in root for "response to ethylene", "regulation of cellular process" and others (Fig. 3B), while in leaf for "photosynthesis", "protein-chromophore linkage" and others (Fig. 3C). According to functional analysis terms, leaves and roots of *M. illicifoia* also differ at levels of molecular function (MF), with transcripts overexpressed in roots being mainly associated with "calcium ion binding", "iron ion binding" and others (Fig. 4B), while the overexpressed leaf transcripts are associated with "oxidoreductase activity", "chlorophyll binding" and others (Fig. 4C).

Significant GO terms linked to secondary methabolism were found in 295 differentially expressed transcripts, 164 in root and 131 in leaf. Some terms were found enriched in specific tissue, for example, "2-oxoglutarate-dependent dioxygenase activity" and "response to herbivore" in root and "beta-amyrin synthase activity" and "triterpenoid biosynthetic process" in leaf. Coincident terms like "oxidoreductase activity" were observed in overexpressed transcripts from both tissues (Table 2).

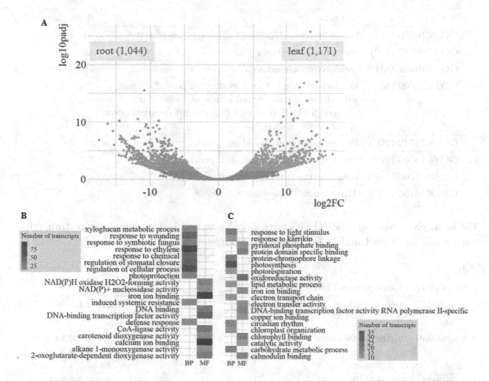

Fig. 4. Gene expression differences between root and leaf tissues of *Maytenus ilicifolia*. A. The values of -$\log_{10}$ adjusted p-value were plotted according to the differential expression between root and leaf ($\log_2$ fold change). Differentially expressed root transcripts are high-lighted in brown (left) and differentially expressed leaf transcripts, in green (right). Top ten most represented terms of gene ontology enrichment analysis in biological process (BP) and molecular function (MF) for differentially expressed transcripts for **B.** root and **C.** and leaf (Color figure online)

The comparative transcriptome analysis led to the identification of 350 and 487 transcripts associated with Enzyme Commission (EC) numbers in root and leaf, respectively. These tissue-biased transcripts were mapped onto the KEGG pathway database for the "Biosynthesis of plant secondary metabolites map" (ko01060) and related pathways. Enzymes involved in "monoterpenoid biosynthesis" and isoflavonoid biosynthesis" were identified in root overexpressed transcripts while "flavonoid biosynthesis" and "Biosynthesis of alkaloids derived from histidine and purine" in leaf (Table 3).

Taking together, the results of GO enrichment analysis and KEGG mapping of transcripts overexpressed in root or leaf of *M. ilicifolia* confirmed the well-

Table 2. Number of transcripts overexpressed in *Maytenus ilicifolia* root or leaf characterized according to enriched GO terms involved in secondary metabolism.

GO ID	GO term	Root	Leaf
GO:0016706 (MF)	2-oxoglutarate-dependent dioxygenase activity	26	–
GO:0080027 (BP)	Response to herbivore	15	–
GO:0016491 (MF)	Oxidoreductase activity	7	25
GO:0016709 (MF)	Oxidoreductase activity, acting on paired donors, with incorporation or reduction of molecular oxygen, NAD(P)H as one donor, and incorporation of one atom of oxygen	6	2
GO:0019742 (BP)	Pentacyclic triterpenoid metabolic process	3	–
GO:0016106 (BP)	Sesquiterpenoid biosynthetic process	3	–
GO:0042300 (MF)	Beta-amyrin synthase activity	–	6
GO:0016104 (BP)	Triterpenoid biosynthetic process	–	5

Table 3. Enzymes mapped KEGG pathways identified in the comparative transcriptome analyses of root and leaf of *Maytenus ilicifolia*.

Root		
map00902	Monoterpenoid biosynthesis	EC:2.1.1.50
map00943	Isoflavonoid biosynthesis	EC:2.1.1.46
Leaf		
map00230	Purine metabolism	EC:2.4.2.14 EC:2.7.6.5 EC:3.5.2.5
map00941	Flavonoid biosynthesis	EC:2.3.1.133
map00950	Isoquinoline alkaloid biosynthesis	EC:1.4.3.21 EC:2.6.1.5
map00960	Tropane, piperidine and pyridine alkaloid biosynthesis	EC:1.4.3.21 EC:2.6.1.5
map01064	Biosynthesis of alkaloids derived from ornithine, lysine and nicotinic acid	EC:1.4.3.21 EC:2.6.1.5
map01065	Biosynthesis of alkaloids derived from histidine and purine	EC:4.1.2.13 EC:1.2.1.9 EC:2.4.2.14 EC:2.7.6.5 EC:5.1.3.1 EC:1.1.1.49 EC:2.2.1.1 EC:3.5.2.5

reported SMs accumulation reveled by other methodological procedures, including flavonoids, triterpenes, and sesquiterpenes in leaves [2], while roots contain terpenes, triterpenes, alkaloids and especially the quinonemethide triterpenes [5,13,14].

Finally, from the present study, an extensive transcriptome dataset has been generated from *de novo* sequencing analyses of *M. ilicifolia*. The coverage of the transcriptome data is consistent to discover genes involved in the secondary metabolic pathways. Therefore, choosing the root and the leaf for comparative transcriptome analysis facilitated the identification of the genes involved in the organ-specific biosynthesis, an approach widely used for mining and identifying novel genes in biosynthesis of SMs in plants[3,6,7,18,25,26].

References

1. Coppede, J.S., et al.: Cell cultures of Maytenus ilicifolia Mart. Are richer sources of quinone-methide triterpenoids than plant roots in natura. Plant Cell Tissue Organ Cult. (PCTOC) **118**(1), 33–43 (2014). 10/f56kq9
2. De Souza, L.M., Cipriani, T.R., Iacomini, M., Gorin, P.A.J., Sassaki, G.L.: HPLC/ESI-MS and NMR analysis of flavonoids and tannins in bioactive extract from leaves of Maytenus ilicifolia. J. Pharm. Biomed. Anal. **47**(1), 59–67 (2008). 10/c4vp7v
3. Devi, K., Mishra, S.K., Sahu, J., Panda, D., Modi, M.K., Sen, P.: Genome wide transcriptome profiling reveals differential gene expression in secondary metabolite pathway of Cymbopogon winterianus OPEN, **6**(21026), 1–11 (2016). 10/f79vzf. Nature Publishing Group
4. Dziggel, C., Schäfer, H., Wink, M.: Tools of pathway reconstruction and production of economically relevant plant secondary metabolites in recombinant microorganisms. Biotechnol. J. **12**(1), 1–14 (2017). 10/f3tn87
5. Filho, W.B., Corsino, J., Bolzani, V.d.S., Furlan, M., Pereira, A.M.S., França, S.C.: Quantitative determination of cytotoxicFriedo-nor-oleanane derivatives from five morphological types of Maytenus ilicifolia (celastraceae) by reverse-phase high-performance liquid chromatography. Phytochem. Anal. Int. J. Plant Chem. Biochem. Tech. **13**(2), 75–78 (2002). https://doi.org/10.1002/PCA.626
6. Guo, D., Kang, K., Wang, P., Li, M., Huang, X.: Transcriptome profiling of spike provides expression features of genes related to terpene biosynthesis in lavender. Sci. Rep. **10**(1), 1–13 (2020). https://doi.org/10.1038/s41598-020-63950-4
7. Hansen, N.L., et al.: The terpene synthase gene family in Tripterygium wilfordii harbors a labdane-type diterpene synthase among the monoterpene synthase TPS-b subfamily. Plant J. **89**(3), 429–441 (2017). 10/f9qghw
8. Hartmann, T.: 10/bvxmg2
9. Jan, R., Asaf, S., Numan, M., Lubna, Kim, K.M.: Plant secondary metabolite biosynthesis and transcriptional regulation in response to biotic and abiotic stress conditions. Agronomy **11**(5), 1–31 (2021). 10/gmk7dd
10. Li, W., et al.: De novo leaf and root transcriptome analysis to explore biosynthetic pathway of Celangulin v in Celastrus angulatus maxim. BMC Genomics **20**(1), 1–15 (2019). 10/gz2c
11. Liu, M.H., et al.: Transcriptome analysis of leaves, roots and flowers of Panax notoginseng identifies genes involved in ginsenoside and alkaloid biosynthesis. BMC Genomics **16**(1), 1–12 (2015). 10/f69r7v

12. Mariot, M.P., Barbieri, R.L.: Metabólitos secundários e propriedades medicinais da espinheira-santa (Maytenus ilicifolia Mart. ex Reiss. e M. aquifolium Mart.). Revista Brasileira de Plantas Medicinais **9**(3), 89–99 (2007)
13. Paz, T.A., et al.: Proteome profiling reveals insights into secondary metabolism in Maytenus ilicifolia (Celastraceae) cell cultures producing quinonemethide triterpenes. Plant Cell Tissue Organ Cult. **130**(2), 405–416 (2017). 10/gbpzrf
14. Périco, L.L., Rodrigues, V.P., de Almeida, L.F.R., Fortuna-Perez, A.P., Vilegas, W., Hiruma-Lima, C.A.: Maytenus ilicifolia Mart. ex Reissek pp. 323–335 (2018). https://doi.org/10.1007/978-94-024-1552-0_29
15. Pradhan, J., Sahoo, S., Lalotra, S., Sarma, R.: Positive impact of abiotic stress on medicinal and aromatic plants. Int. J. Plant Sci. **12**(2), 309–313 (2017). 10/gz2g
16. Ramakrishna, A., Ravishankar, G.A.: Influence of abiotic stress signals on secondary metabolites in plants. Plant Signal. Behav. **6**(11), 1720–1731 (2011). 10/fx4rjw
17. Saddique, M., Kamran, M., Shahbaz, M.: Differential Responses of Plants to Biotic Stress and the Role of Metabolites. Elsevier Inc. (2018). 10/gz2h
18. Upadhyay, S., Phukan, U.J., Mishra, S., Shukla, R.K.: De novo leaf and root transcriptome analysis identified novel genes involved in Steroidal sapogenin biosynthesis in Asparagus racemosus. BMC Genomics **15**(1), 1–13 (2014). 10/gb3gr4
19. Van Loon, L.C., Rep, M., Pieterse, C.M.: Significance of inducible defense-related proteins in infected plants. Ann. Rev. Phytopathol. **44**, 135–162 (2006). 10/csvjsr
20. Vellosa, J.C., et al.: Antioxidant activity of Maytenus ilicifolia root bark. Fitoterapia **77**(3), 243–244 (2006). 10/dzm4t9
21. Wink, M.: Introduction: biochemistry, physiology and ecological functions of secondary metabolites. Biochem. Plant Second. Metab. Second Ed. **40**, 1–19 (2010). 10/b8sdms
22. Wink, M.: Secondary metabolites: deterring herbivores. In: Encyclopedia of Life Sciences, pp. 1–9, March 2010. 10/c65zd8
23. Wink, M., Schimmer, O.: Molecular modes of action of defensive secondary metabolites, vol. 39 (2010). 10/cpz4j7
24. Yang, L., Wen, K.S., Ruan, X., Zhao, Y.X., Wei, F., Wang, Q.: Response of plant secondary metabolites to environmental factors. Molecules **23**(4), 1–26 (2018). 10/gdrnqc
25. Younesi-Melerdi, E., Nematzadeh, G.A., Pakdin-Parizi, A., Bakhtiarizadeh, M.R., Motahari, S.A.: De novo RNA sequencing analysis of Aeluropus littoralis halophyte plant under salinity stress. Sci. Rep. **10**(1), 1–14 (2020). 10/gz2m
26. Zhang, C., Yao, X., Ren, H., Chang, J., Wang, K.: RNA-Seq reveals flavonoid biosynthesis-related genes in pecan (Carya illinoinensis) kernels. J. Agric. Food Chem. **67**, 148–158 (2018). 10.gz2n

Hypusine Plays a Role in the Translation of Short mRNAs and Mediates the Polyamine and Autophagy Pathways in *Saccharomyces Cerevisiae*

Ana Carolina Silva Paiva(✉) (iD) , Fernanda Manaia Demarqui(iD),
Mariana Marchi Santoni(iD), Sandro Roberto Valentini(iD),
and Cleslei Fernando Zanelli(iD)

Department of Biological Sciences, School of Pharmaceutical Sciences,
São Paulo State University (UNESP), Araraquara, SP 14800-903, Brazil

Abstract. The cell highly regulates the translational process aiming to maintain cellular stability and viability for mechanisms at the transcriptional, translational, or metabolic level, such as the control of transcripts forwarded for translation or autophagy. The translation elongation factor 5A (eIF5A) is evolutionarily conserved and essential in eukaryotic cells. eIF5A undergoes a post-translational modification, called hypusination, which has two enzymatic steps. The first stage, catalyzed by the deoxyhypusine synthase, occurs in a spermidine-dependent manner. Spermidine is a polyamine in which intracellular imbalance can affect some cellular processes. Studies show that this modification is fundamental to the role of eIF5A in the cell, assisting in the translation of a subset of mRNA. We analyzed transcriptional and translational profiles of the deoxyhypusine synthase mutant (*dys1-1*) in *Saccharomyces cerevisiae*. From Polysome-seq, our results showed that the lack of hypusination leads to the impairment on the translation of short ORFs, that code ribosomal mitochondrial proteins. From both profiles, the expression of genes and transcription factors of the polyamine pathway, which needs strict cell control, was altered. Besides, the inhibition of hypusination by GC7 showed an increase in the protein level of two autophagy proteins, Atg1 and Atg33, the latter is specific to mitophagy. In response to the metabolic problems caused by non-hypusination, the cell can respond with mitophagy and macroautophagy to maintain cell stability.

Keywords: Deoxyhypusine synthase · eIF5A · GC7.

1 Introduction

Translation comprises protein synthesis, where ribosomes and translational factors recognize and decode the messenger RNA template by cycling through

This study was financially supported by grant 2010/50044-6, São Paulo Research Foundation (FAPESP) and Coordenação de Aperfeiçoamento de Pessoal de Nível Superior - Brasil (CAPES) - Finance Code 001.

© Springer Nature Switzerland AG 2021
P. F. Stadler et al. (Eds.): BSB 2021, LNBI 13063, pp. 15–25, 2021.
https://doi.org/10.1007/978-3-030-91814-9_2

translation initiation, elongation, termination phases and ribosome recycling [29,32]. The translation elongation factor 5A (eIF5A), a structural ortholog of the bacterial EF-P protein, is evolutionarily conserved among eukaryotes (eIF5A) and archeas (aIF5A). eIF5A undergoes hypusination, an exclusive and spermidine-dependent post-translational modification, that converts a specific lysine residue (K51 in *Saccharomyces cerevisiae*) by the action of two enzymes: deoxyhypusine synthase (Dys1 in yeast) and deoxyhypusine hydroxylase (Lia1 in yeast) [6,9,27,28]. The enzymes involved in this exclusive modification are also conserved [24]. eIF5A is high abundant and essential for cell viability in eukaryotic cells. Hypusinated eIF5A (eIF5AH) promotes a stabilization of the peptide bond during the elongation of a subset of mRNAs, caracterized by low translational processability, and is also related to termination [30].

Spermidine, an endogenous polyamine involved in hypusination, has also been reported to induce autophagy [19]. Additionally, polyamines have roles in cellular proliferation, DNA binding, ion channels modulation and protein synthesis [9,21,25]. A structural analogue of spermidine, N1-guanyl-1,7-diaminoheptane (GC7), acts as an inhibitor of the first hypusination enzyme, anchoring at the spermidine-binding site of Dys1 [2].

In this study, we combined polysome profiling and next-generation sequencing as a measure of translational profile to investigate the role of hypusination in global translation in yeast. This characterization of the translational profile was revealed to be correlated to the already consolidated ribosomal profile [15]. We used the deoxyhypusine synthase mutant (*dys1-1*) datasets of *Saccharomyces cerevisiae* [8] to identify transcripts dependent on eIF5A hypusination at transcriptional and translational levels. The *dys1-1* mutant displays compromised translation of specific functional groups, including mitochondrial ribosomal proteins. Besides, we found the mutation generated expression changes in genes envolved in polyamine pathway, which possibly mimics the effect caused by the addition of spermidine in the cell and consequently undergoing autophagy. Additionally, non-hypusination due to GC7 leads to an increase in the level of proteins related to macroautophagy and mitophagy.

2   Materials and Methods

2.1   RNA-seq Data Analysis

RNA-seq and Polysome-seq datasets for *DYS1* and *dys1-1* were taken from [8] and the *Saccharomyces cerevisiae* R64-1-1 S288C reference genome was used for the analysis. Bioinformatic analysis was conducted according to [8]. Briefly, adaptor sequence was trimmed using Trim-Galore! (Galaxy Tool Version: 0.4.3.1 + galaxy1) and° low-quality reads (Phred score <25) were discarded. First, trimmed reads were alignment to the RNA gene database FASTA file to remove noncoding RNAs using Bowtie software (Galaxy Tool Version: 1.1.2) with the parameters -v 2 -y -a -m 1 -best -strata -S -p 4. The remaining reads were then aligned to the genome using Stringtie software (Galaxy Tool Version: 1.3.4) with standard parameters. The mapped reads were normalized TMM and differentially expressed genes were identified using anota2seq [22]. Significance was determined using an adjusted p-value limit of 0.05.

Translational efficiency (TE) was defined by transcript abundance in the translatome profiling to the abundance of the respective transcript in the transcriptome profiling. Changes in translational efficiency - changes for *dys1-1* strain compared to *DYS1* strain in translated mRNA after adjustment for corresponding changes in total RNA, also known as ΔTE - was calculated by anota2seq using Analysis of Partial Variance (APV).

Gene ontology (GO) terms and biosynthetic pathways of differentially expressed genes were determined using Yeastmine database [10], considering p-value <0.05 for the Holm-Bonferroni correction test. Fisher exact test was used to test for statistically significant differences, and the Holm-Bonferroni correction test procedure to adjust for the effects of multiple tests [3]. GO terms were considered significant when FDR <0.01.

2.2 Strain and Growth Conditions

Saccharomyces cerevisiae strains SVL613 (MATa leu2 trp1 ura3 his3 dys1::HIS3 [*DYS1*/TRP1/CEN - pSV520]) and SVL614 (MATa leu2 trp1 ura3 his3 *dys1*::HIS3 [*dys1-1*/TRP1/CEN - pSV730]) were used to qPCR experiments. The mutant *dys1-1* (*dys1-1*W75R,T118A,A147T) and its special growth conditions are described in [11].

For western blot assays, strains MATa ATG1-TAP::HIS3 his3Δ1 leu2Δ0 ura3Δ0 met15Δ0 and MATa ATG33-TAP::HIS3 his3Δ1 leu2Δ0 ura3Δ0 met15Δ0, from TAP-tagged yeast collection, were used [33]. The procedures for cell growth were performed according to standard protocols [1]. Cells were grown to OD600 $\sim$ 0.4 and then treated with 1 mM N1-guanyl-1,7-diaminoheptane (GC7; Biosearch Technologies) or vehicle control (0.1 mM acetic acid) in culture medium for 12 h at 25 °C. For GC7 treatment, 1 mM aminoguanidine was added to prevent degradation by monoamine oxidases [31]. The cultures were centrifuged, and the cell pellets were stored at −80 °C.

2.3 RNA Isolation and qRT-PCR

For total RNA isolation, SVL613 and SVL614 strains were grown in exponential phase an OD $\sim$ 0.6. Cultures were centrifuged and cell pellets were stored at −80 °C. Cell lysis was conducted with zymolyase and total RNA was extracted using the RNeasy mini kit (cat. number 74104, Qiagen). Total RNA was quantified using a NanoDrop 2000 Spectrophotometer (ThermoFisher) and its integrity were verified by electrophoresis gel on 2100 Bioanalyzer equipment (Agilent, Santa Clara, CA), using a High Sensitivity Total RNA Analysis Chip.

For the RNA analysis by real-time PCR, 5000 ng of RNA from SVL613 and SVL614 strains was treated with DNaseI kit (Sigma - AMPD1-1KT) and the first-strand cDNA was synthesized by the SuperScript® IV Reverse Transcriptase (RT) kit (Life Technologies) following the manufacturer's instructions.

Real-time PCR was performed in a 7500 Real-Time PCR instrument (Applied Biosystems) with the Power SYBR^{TM} Green PCR Master Mix detection system (Life Technologies). Reactions were performed in a total 20 µl volume with 10 ug of the synthesized cDNA. Each sample was analyzed in triplicate with independent biological replicates per sample. Values were normalized to the steady-state SCR1 mRNA levels using ddCt method [17].

2.4   Protein Extraction and Western Blot Analysis

Protein extracts from TAP-tagged yeast collection strains were obtained from mechanical lysis of the cell pellets vortexed for 12 min at 4 °C (alternating every 3 min, 1 min on ice) in lysis buffer (200 mM Tris-HCl, pH 7.5; 2 mM dithiothreitol; 2 mM EDTA, pH 8; 0.2% Triton X-100) with protease inhibitors (5 µg.mL-1 of pepstatin, leupeptin, aprotinin and chymostatin; and 2 Mm PMSF) and glass beads. The cell extract was centrifuged (10,600 xg, at 4 °C, 15 min), the supernatant was removed, and the total protein concentration was determined by the Lowry method [23].

A specific amount of protein (40 µg for Atg1 and 25 µg for Atg33 western blots) in 6X SDS loading buffer (0.3 M Tris-HCl; 0.6 M DTT; 10% SDS; 0.06% Bromophenol blue; 30% glycerol) was heated in 96 °C heating block for 5 min. Protein samples were analyzed by SDS-PAGE using 12% polyacrylamide gels and transferred to nitrocellulose membranes which were blocked (10% nonfat powdered milk; PBS 1X; 0.25% Tween-20) and incubated with antibodies. Antibody dilutions were as follows: anti-TAP (Sigma-Aldrich P1291), 1:4,000; a rabbit polyclonal anti-Rpl5 antibody (yeast), 1:20,000; a rabbit polyclonal anti-hypusine antibody (Merck Millipore ABS1064) 1:2,500; and anti-rabbit secondary antibody (Sigma-Aldrich A9169), 1:20,000. Immunoreactivity protein signals were quantified using ImageJ software.

3   Results and Discussion

3.1   Translation of Short ORFs is Impaired in *dys1-1* Mutant

A polysome consists in an mRNA occupied by two or more ribosomes and the number of ribosomes in an ORF relates directly to its length [13,14]. We found significantly different ORF length distribution (p-value < 0,001 for Mann-Whitney test) for up and down regulated mRNAs in *dys1-1* translational profile, compared to the *DYS1*, implying that ORFs recruited for translation in the mutant strain are, on average, longer (median: 1152 nt), than those for the wild type (1482 nt), while the translation of short ORFs (720 nt) was strongly affected (Fig. 1A). These results were not observed for transcriptional profiling (Fig. 1B).

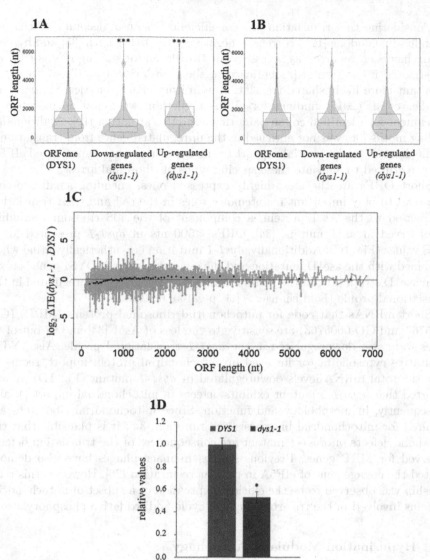

Fig. 1. Hypusination interferes with the gene expression of short ORFs. Violin and blox plot of ORF length in the (A) translational and (B) transcriptional profiles. [***] indicates p-value <0,001 for Mann-Whitney test. (C) Relationship between ORF length and translation efficiency (TE) changes in *dys1-1* mutant. The values shown represent the average percent change in TE for bins of 100 genes arranged by length. The ORF lengths shown correspond to the point at which the average ORF length of the bin exceeds the indicated value. Shaded areas represent absolute values for all genes. (D) The relative quantification of mRNA encoding YDJ1 using qRT-PCR. The qRT-PCR was carried out using primers specific for YDJ1 and SCR1 (control). Error bar represents standard deviation from 3 different biological replicates, tested by Mann-Whitney, using a p-value limit of 0.05.

Considering that translation process efficiency - sum of decoding efficiencies for individual codons [12] - correlates negatively to ORF length [35] and hypusination has been reported as necessary for translation of stalling motifs [30], we investigated if the transcripts enriched in the translational profile of the *dys1-1* mutant more likely shared translation inhibitory characteristics. Similarly, as Schuller et al. (2017) findings for eIF5A depletion, we found no evidence of enrichment for identified codon pairs that inhibit translation [12]. Additionally, stalling motifs [30] was not enriched in the upregulated genes from translational profile dataset, lending initial support to the hypothesis that hypusinated eIF5A is not recruited to alleviate these specific types of ribosomal arrest.

Short ORFs are the most highly expressed ones, encoding small proteins that tend to play important maintenance roles in the cell and their translation are related to the Asc1 protein, a component of the 40S ribosomal subunit. As observed in *asc1* mutant [34], ORFs <500 nts in *dys1-1* presented lower ΔTE values (Fig. 1C). Additionally, *dys1-1* mutation is synthetically lethal when combined with the asc1D mutation and the overexpression of DYS1 gene is toxic to an asc1D strain [11]. In the present study, ASC1 gene is downregulated in the translational profile (Fold change: -1.1, p-value: 0.004).

Short mRNAs that code for mitochondrial ribosomal proteins (MRP) (GO 0005762 and GO 0005763) are sensitive to the loss of Asc1 [34] and 67 out of 84 genes were also downregulated in the *dys1-1* translational profile. Also, Ydj1, a putative cytosolic factor involved in mitochondrial protein import, seems to have the total mRNA levels downregulated in *dys1-1* mutant (Fig. 1D). It was reported that a *ydj1*Δ mutant exhibits defects in mitochondrial import [4] and consequently, in morphology and function. Since mitochondrial ribosomes are required for mitochondrial biogenesis and function [34], it is plausible that the metabolic defects on *dys1-1* mutant are consequences of the translation defects observed for MRP genes. Previous studies in macrophages have also demonstrated the engagement of eIF5A in cellular respiration [26]. However, this relationship was observed to be the efficient expression of a subset of mitochondrial proteins involved in the tricarboxylic acid cycle and oxidative phosphorylation.

3.2   Hypusination Modulates Autophagy

We concerned if the decrease in Dys1 protein levels and consequent reduction in the amount of hypusine-containing eIF5A in *dys1-1* mutant [11] can affect the cell metabolism due to polyamines accumulation, similar to the result observed in mammalian cells by spermidine addition [16]. A transcriptional profile study with the double mutant $\Delta spe3$ $\Delta fms1$ strain (spermidine auxotroph), treated with spermidine in excess, showed significant changes in biosynthesis of methionine, arginine, lysine, NAD, and biotin [5]. In our large-scale tracking, we observed that genes involved in these pathways were transcriptionally upregulated in the *dys1-1* mutant.

Specifically for arginine biosynthesis, a polyamine precursor, we observed an enrichment of this pathway, through Yeastmine database. When it comes

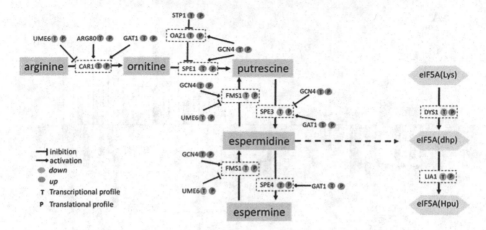

Fig. 2. Representation of the polyamine biosynthetic pathway and hypusination. The levels of expression of the genes are represented for both large scale profiles. The transcription factors of the polyamine pathway that had their expression altered in the *dys1-1* mutant are also indicated.

to the polyamine pathway, most of the genes and their respective transcription factors have changed transcriptional and translational levels in the *dys1-1* mutant (Fig. 2). These results suggest that *dys1-1* mutant generate an imbalance of polyamines levels. Those polyamines are essential for many basic cellular functions; their intracellular levels are tightly regulated in their biosynthesis, catabolism and/or transport. So, polyamine intracellular imbalance could affect some cellular processes such as tran-scription, translation, gene expression regulation, stress resistance and autophagy [16,21].

In fact, we observed that most proteins involved in autophagy processes (Atgs) have an increased translational efficiency in *dys1-1* when compared to DYS1 (Fig. 3A), configuring an increase in these mRNAs recruited for translation. Additionally, western blot analysis of two Atgs (Atg1 and Atg33) in wild type strain with loss of hypusination caused by treatment with GC7, showed an increase in the protein level for both proteins (Fig. 3B and 3C). The Atg1 protein acts in the regulation and signaling to induce the formation of autophagosomes [7,36] and Atg33 is a specific mitophagy protein. GC7 treatment also attenuated anoxia-induced generation of reactive oxygen species in these cells and in normoxic conditions and decreased the mitochondrial oxygen consumption rate of cultured cells and mice [20]. Based on the above considerations polyamines and eIF5A can modulate autophagy and hypusination could be the key of this relationship [18,19].

Finally, our data suggest that hypusination supports the expression of a set of short proteins, including mitochondrial ribosome proteins, providing a possible explanation to why lack of hypusination results in cell respiration defect.

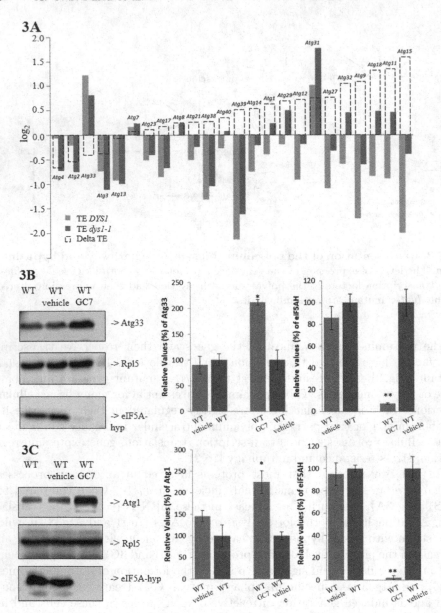

Fig. 3. Hypusination modulates autophagy and mitophagy. (A) Bar graph represents the rate of translation efficiency (TE - number of reads translational profile/number of reads from the transcriptional profile) of the wild and *dys1-1* mutant, as well as the delta TE (*dys1-1*/WT) for the ORFs of the Atgs proteins. Western blot of the wild type strain of the TAP collection without/with the addition of GC7 to identify the protein level of (B) Atg33 and (C) Atg1, and their respective relative quantifications (%). To measure the intensity of the bands, the Image-J software was used, and the Rpl5 protein as a normalizer. Values are shown as mean ± SEM, n = 2. [*] p-value <0.05 [**] p-value <0.01 for t-test. WT: wild type; WT vehicle: wild type treated with 0.1 mM acetic acid (GC7 vehicle); WT GC7: wild type treated with 1 mM of GC7 plus 1 mM of aminoguanidine.

This is a consequence of translational problems due to the lack of translational processivity, as well as to the cellular imbalance of polyamines, resulting in long ORFs occupying ribosomes. As a response to maintain cellular homeostasis, there is an induction of macroautophagy and mitophagy.

References

1. Abelson, J.N., Simon, M.I., Guthrie, C., Fink, G.R.: Guide to yeast genetics and molecular biology **194**, 1–863 (1991). https://doi.org/10.2307/3760517
2. Annette, K., et al.: Modification of eukaryotic initiation factor 5A from plasmodium vivax by a truncated deoxyhypusine synthase from plasmodium falciparum: an enzyme with dual enzymatic properties. Bioorg. Med. Chem. **15**(18), 6200–6207 (2007). https://doi.org/10.1016/J.BMC.2007.06.026
3. Benjamini, Y., Yekutieli, D.: The control of the false discovery rate in multiple testing under dependency. Ann. Stat. **29**, 1165–1188 (2001). https://doi.org/10.1214/aos/1013699998
4. Caplan, A.J., Cyr, D.M., Douglas, M.G.: YDJ1p facilitates polypeptide translocation across different intracellular membranes by a conserved mechanism. Cell **71**(7), 1143–1155 (1992). https://doi.org/10.1016/S0092-8674(05)80063-7
5. Chattopadhyay, M.K., Chen, W., Poy, G., Cam, M., Stiles, D., Tabor, H.: Microarray studies on the genes responsive to the addition of spermidine or spermine to a saccharomyces cerevisiae spermidine synthase mutant. Yeast **26**(10), 531–544 (2009). https://doi.org/10.1002/yea.1703
6. Chen, K.Y., Liu, A.Y.: Biochemistry and function of hypusine formation on eukaryotic initiation factor 5A. NeuroSignals **6**, 105–109 (1997). https://doi.org/10.1159/000109115
7. Chen, Y., Klionsky, D.J.: The regulation of autophagy - unanswered questions (2011). https://doi.org/10.1242/jcs.064576
8. Demarqui, F.M., Paiva, A.C.S., Santoni, MMi., Watanabe, T.F., Valentini, S.R., Zanelli, C.F.: Polysome-seq as a measure of translational profile from deoxyhypusine synthase mutant in *saccharomyces cerevisiae*. In: BSB 2020. LNCS, vol. 12558, pp. 168–179. Springer, Cham (2020). https://doi.org/10.1007/978-3-030-65775-8_16
9. Dever, T.E., Dinman, J.D., Green, R.: Translation Elongation and Recoding in Eukaryotes. Cold Spring Harb. Perspect Biol. **10**, a032649 (2018). https://doi.org/10.1101/cshperspect.a032649
10. Engel, S.R., et al.: The reference genome sequence of saccharomyces cerevisiae: then and now. G3: Genes Genomes Genet **4**(3), 389–398 (2014). https://doi.org/10.1534/g3.113.008995
11. Galvão, F.C., Rossi, D., Silveira, W.D.S., Valentini, S.R., Zanelli, C.F.: The deoxyhypusine synthase mutant dys1-1 reveals the association of eIF5A and Asc1 with cell wall integrity. PLOS ONE **8**(4), e60140 (2013). https://doi.org/10.1371/JOURNAL.PONE.0060140
12. Gamble, C.E., Brule, C.E., Dean, K.M., Fields, S., Grayhack, E.J.: Adjacent codons act in concert to modulate translation efficiency in yeast. Cell **166**(3), 679–690 (2016). https://doi.org/10.1016/j.cell.2016.05.070
13. Heyer, E.E., Moore, M.J.: Redefining the translational status of 80s monosomes. Cell **164**(4), 757–769 (2016). https://doi.org/10.1016/j.cell.2016.01.003, https://www.sciencedirect.com/science/article/pii/S0092867416000040

14. Hurowitz, E.H., Brown, P.O.: Genome-wide analysis of mRNA lengths in saccharomyces cerevisiae. Genome Biol. **5**(1), 3889–3894 (2003). https://doi.org/10.1186/gb-2003-5-1-r2

15. Ingolia, N.T., Ghaemmaghami, S., Newman, J.R., Weissman, J.S.: Genome-wide analysis in vivo of translation with nucleotide resolution using ribosome profiling. Science **324**(5924), 218–223 (2009). https://doi.org/10.1126/science.1168978

16. Ivanov, I.P., et al.: Polyamine control of translation elongation regulates start site selection on antizyme inhibitor mRNA via ribosome queuing. Mol. Cell **70**(2), 254-264.e6 (2018). https://doi.org/10.1016/j.molcel.2018.03.015

17. KJ, L., TD, S.: Analysis of relative gene expression data using real-time quantitative PCR and the 2(-delta delta C(T)) method. Methods (San Diego, Calif.) **25**(4), 402–408 (2001). https://doi.org/10.1006/METH.2001.1262

18. Lubas, M., et al.: eIF 5A is required for autophagy by mediating ATG 3 translation. EMBO Rep. **19**(6), e46072 (2018). https://doi.org/10.15252/embr.201846072

19. Madeo, F., Tobias, E., Sabrina, B., Christoph, R., Guido, K.: Spermidine: a novel autophagy inducer and longevity elixir. Autophagy **6**(1), 160–162 (2010). https://doi.org/10.4161/AUTO.6.1.10600

20. Melis, N., et al.: Targeting eIF5A hypusination prevents anoxic cell death through mitochondrial silencing and improves kidney transplant outcome. J. Am. Soc. Nephrol. **28**(3), 811–822 (2017). https://doi.org/10.1681/ASN.2016010012

21. Miller-Fleming, L., Olin-Sandoval, V., Campbell, K., Ralser, M.: Remaining mysteries of molecular biology: the role of polyamines in the cell (2015). https://doi.org/10.1016/j.jmb.2015.06.020

22. Oertlin, C., Lorent, J., Murie, C., Furic, L., Topisirovic,I., Larsson, O.: Generally applicable transcriptome-wide analysis of translation using anota2seq. Nucleic Acids Res. **47**(12), e70 (2019). https://doi.org/10.1093/nar/gkz223

23. Oliver H, L., Nira J, R., A Lewis, F., Rose J, R.: Protein measurement with the folin phenol reagent. J. Biol. Chem. **193**(1), 265–275 (1951)

24. Park, M.H., Nishimura, K., Zanelli, C.F., Valentini, S.R.: Functional significance of eIF5A and its hypusine modification in eukaryotes. Amino Acids **38**(2), 491–500 (2010). https://doi.org/10.1007/s00726-009-0408-7

25. Pegg, A.E.: Functions of polyamines in mammals (2016). https://doi.org/10.1074/jbc.R116.731661

26. Puleston, D.J., et al.: Polyamines and eIF5A hypusination modulate mitochondrial respiration and macrophage activation. Cell Metab. **30**(2), 352-363.e8 (2019). https://doi.org/10.1016/j.cmet.2019.05.003

27. Rossi, D., Kuroshu, R., Zanelli, C.F., Valentini, S.R.: eIF5A and EF-P: two unique translation factors are now traveling the same road (2014). https://doi.org/10.1002/wrna.1211

28. Schnier, J., Schwelberger, H.G., Smit-McBride, Z., Kang, H.A., Hershey, J.W.: Translation initiation factor 5A and its hypusine modification are essential for cell viability in the yeast Saccharomyces cerevisiae. Mol. Cell. Biol. (1991). https://doi.org/10.1128/MCB.11.6.3105

29. Schuller, A.P., Green, R.: Roadblocks and resolutions in eukaryotic translation. Nat. Rev. Mol. Cell Biol. **19**, 526–541 (2018). https://doi.org/10.1038/s41580-018-0011-4

30. Schuller, A.P., Wu, C.C.C., Dever, T.E., Buskirk, A.R., Green, R.: eIF5A functions globally in translation elongation and termination. Mol. Cell **66**(2), 194-205.e5 (2017). https://doi.org/10.1016/j.molcel.2017.03.003

31. Shi, X.P., Yin, K.C., Ahern, J., Davis, L.J., Stern, A.M., Waxman, L.: Effects of N1-guanyl-1,7-diaminoheptane, an inhibitor of deoxyhypusine synthase, on the growth of tumorigenic cell lines in culture. Biochim. et Biophys. Acta - Mol. Cell Res. **1310**(1), 119–126 (1996). https://doi.org/10.1016/0167-4889(95)00165-4
32. Shirokikh, N.E., Preiss, T.: Translation initiation by cap-dependent ribosome recruitment: Recent insights and open questions. Wiley Interdisc. Rev. RNA **9**(4),(2018). https://doi.org/10.1002/wrna.1473
33. Sina, G., et al.: Global analysis of protein expression in yeast. Nature **425**(6959), 737–741 (2003). https://doi.org/10.1038/NATURE02046
34. Thompson, M.K., Rojas-Duran, M.F., Gangaramani, P., Gilbert, W.V.: The ribosomal protein Asc1/RACK1 is required for efficient translation of short mRNAs. eLife **5**, e11154 (2016). https://doi.org/10.7554/eLife.11154
35. Weinberg, D.E., Shah, P., Eichhorn, S.W., Hussmann, J.A., Plotkin, J.B., Bartel, D.P.: Improved ribosome-footprint and mRNA measurements provide insights into dynamics and regulation of yeast translation. Cell Rep. **14**(7), 1787–1799 (2016). https://doi.org/10.1016/j.celrep.2016.01.043
36. Yi, C., et al.: Formation of a Snf1-Mec1-Atg1 module on mitochondria governs energy deprivation-induced autophagy by regulating mitochondrial respiration. Dev. Cell **41**(1), 59-71.e4 (2017). https://doi.org/10.1016/j.devcel.2017.03.007

Topological Characterization of Cancer Driver Genes Using Reactome Super Pathways Networks

Rodrigo Henrique Ramos[1,2](✉) (iD), Jorge Francisco Cutigi[1,2](iD),
Cynthia de Oliveira Lage Ferreira[2](iD), and Adenilso Simao[2](iD)

[1] Federal Institute of Sao Paulo, Sao Carlos, SP, Brazil
ramos@ifsp.edu.br
[2] University of Sao Paulo, Sao Carlos, SP, Brazil

Abstract. Cancer is a complex disease caused by genetic mutations categorized into two groups: passenger and driver. Contrary to passenger, drivers mutations directly impact oncogenesis. Drivers identification is a challenge in cancer genomics, frequently supported by statistical and computational methods. These methods utilize the increasing volume of molecular data related to cancer, gene interactions networks, and pathways. Reactome recently defined 26 Super Pathways that group genes responsible for essential biological processes. Pathways networks carry topological information relative to their biological functions that emerge from genes interactions. Since some pathways are more associated with cancer than others and all have distinct structures, this work aims to characterize cancer driver genes' topological role in Super Pathways networks. We combine data from three different databases to create Super Pathways networks enriched with cancer driver genes information. Results show that Super Pathways networks have distinct topologies and particular roles for drivers. Drivers have significant differences in clustering, betweenness, and closeness centralities when compared to others genes. Attacks using random and intentional removal reveal a remarkable resilience in some Super Pathways networks. Attacks also reveal that drivers in the Programmed Cell Death pathway are more critical than hubs in keeping the network integrity. These distinguishable patterns associated with drivers can support the task of identifying and validate unknown drivers. In addition, recognize the topological role of drivers helps understand the impact mutations in these genes have on pathways structure.

Keywords: Cancer bioinformatics · Cancer drivers genes · Pathways · Protein interaction networks · Complex networks · Topological analysis

© Springer Nature Switzerland AG 2021
P. F. Stadler et al. (Eds.): BSB 2021, LNBI 13063, pp. 26–37, 2021.
https://doi.org/10.1007/978-3-030-91814-9_3

1   Introduction

Cancer is a molecular disease caused by genetic mutations that leads to cells uncontrolled growth and division. Tumors undergo a large number of mutations, but only a small portion of them contributes to oncogenesis and tumor progression. In this context, cancer mutations are classified into two groups: passenger or driver. Contrary to passenger, driver mutations have a direct impact on oncogenesis, conferring a growth advantage to the cell [25].

The study of driver mutations and their associated genes can contribute to understanding the onset and evolution of the disease. Over the past few decades, advances in DNA sequencing have generated several databases specialized in cancer genomics, like TCGA (The Cancer Genome Atlas). These databases allow the development of computational approaches, which include, for instance, the study of how these drivers' genetic changes are commonly involved in different types of cancer [16]. Some recent databases, such as NCG [22], and IntOGen [12], publicly make available online[1] sets of known or predicted cancer driver genes.

Genes are known to interact with each other in common biological functions. Such interactions are part of a complex system that characterizes cell functioning since it comprises many individual parts that work together to emerge biological functions. This cell complex system is commonly modeled as a complex network [14], such as REACTOME FI, in which genes (nodes) and interactions (links) are characterized by non-trivial topologies observed in real phenomena. These networks are frequently used in knowledge-based studies for the detection of driver genes [16].

Pathways are a subset of genes that interact to perform specific biological functions. Each pathway works as cell building blocks and describes the main functions of this complex system [10]. Reactome[2] is a consolidated pathway database. In its most recent paper [8], the authors present the concept of Super Pathways, which represent 26 biological functions that group 1803 sub-pathways. Several studies and analyses involving pathways consider the modeling of pathways as complex networks. This approach is an improvement over other methods of pathway analysis, since it makes possible the topological study of genes, combining their biological function with the topological role in the network [5].

The study of topological characteristics of genes in the network and pathways is an important topic, once it can contribute to understanding the role of drivers and their genes in the networks. The work of [3] shows the gene network centrality measures increase the potential of detecting possible drivers and false drivers. Furthermore, a great number of network-based methods use information about networks to identify significant genes in cancer [18].

In this context, this work models Super Pathways as complex networks to observe the topological characteristics of driver genes and their central role in

---

[1] http://ncg.kcl.ac.uk/ and https://www.intogen.org/.
[2] https://reactome.org/.

such networks, aiming to investigate the hypothesis that driver genes are topologically different from other genes in the same pathway.

In the next section, we describe how data from three different databases were merged. Next, in Sect. 3, we studied the centrality role of drivers in its impact on the resilience of Super Pathways networks. Such studies show that driver nodes have a different topological role than other nodes and some networks are high resilience. Finally, Sect. 4 discusses the results and concludes with possible applications and future work.

2   Method

This section describes the steps taken in the development of this research. Figure 1 shows an overview used to create the Super Pathway networks enriched with cancer driver genes and posterior analyses. The data came from three different databases and are merged. Sections 2.1, 2.2, and 2.3 detail these steps. Next, Sect. 3 discusses the analyses and results.

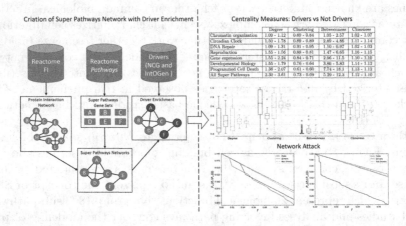

Fig. 1. Approach overview: data pipeline and analyses. We merge data from three different bases to created enriched networks. The cancer driver genes in these networks are analyzed using centrality measures and attacks.

2.1   Reactome Functional Iteration

Reactome is an open-source, open access, manually curated, and peer-reviewed pathway database[3]. Reactome website provides many online Bioinformatics tools to analyze and visualize pathway-related data. One of its tools is ReactomeFIViz that is used to find pathways and network patterns related to cancer, and other types of diseases [26]. ReactomeFIViz is a visualization enrichment tool built upon the Reactome Functional Interaction (FI) network. This network was initially created by merging interactions extracted from human curated pathways

---

[3] https://reactome.org/what-is-reactome.

with interactions predicted using a machine learning approach [27]. Reactome FI last version is from 2020 and has 14006 nodes and 259151 links. The average degree is 37 while the biggest node has 1117 links. This scale-free distribution is characterized by an exponent of 3.8.

2.2 Super Pathways as Reactome FI Sub-networks

Pathways are fundamental units essential to comprehend the emergence of cellular phenomenons [5,10]. Albeit the concept of pathway date back from the 1950s, only in the mid-1990s the firsts pathways databases were created [5]. The association of gene groups to biological function enabled the development of many function enrichment tools. These tools search pathways databases for statistically significant pathways associated with the input gene set given by the user [5]. Recently function enrichment tools no longer consider the pathway as a set of independent genes but as a network of interacting genes. In this sense, pathways network topology analyzes is an evolution over the previous approaches since its combines biological and topological information [5].

Reactome is a consolidated pathway database that consistently updates the genes' interactions and analyzes tools. In the most recent paper [8], they present the concept of Super Pathways, which represent 26 biological functions that group 1803 sub-pathways. We extract induced sub-networks from Reactome FI, creating Super Pathways networks for each Super Pathways genes set. In this paper, we analyze seven Super Pathway Networks with the most percentual presence of cancer driver genes. We created a consensus Super Pathway Network, named All Super Pathways, to represent all the 26 Super Pathways genes sets. Table 1 presents a summary of the chosen Super Pathways and the corresponding Super Pathway Network. LenSet, which is the size of Super Pathway gene set; LenCC, which is the size of the resulting nodes (genes) in the induced network that may have more than one Connect Component (CC); LenLCC, which is the size of nodes in the Largest CC; Driver LCC, which is the number of cancer driver genes in the LCC; Driver %, which shows the percentual presence of drivers in the LCC.

Table 1. Chosen super pathways

Super pathway name	LenSet	LenCC	LenLCC	Driver LCC	Drivers %
Chromatin organization	240	218 (91%)	206 (86%)	45	22
Circadian clock	70	69 (99%)	64 (91%)	12	19
DNA repair	312	290 (93%)	284 (91%)	48	17
Reproduction	114	95 (83%)	81 (71%)	14	17
Gene expression	1536	1392 (91%)	1367 (89%)	194	14
Developmental biology	1097	972 (89%)	962 (88%)	137	14
Programmed cell death	216	208 (96%)	201 (93%)	27	13
All super pathways	11375	9472 (83%)	9399 (83%)	649	7

The seven chosen Super Pathways have specific biological functions and are associated with cancer. **Chromatin organization** refers to the composition and conformation of complexes between DNA, protein, and RNA [20], and have been reported to have a significant influence on regional mutation rates in human cancer cells [23]. **Circadian Clock** is a master regulator of mammalian physiology, regulating daily oscillations of crucial biological processes and behaviors. Notably, circadian disruption has recently been identified as an independent risk factor for cancer and classified as a carcinogen [24]. **DNA Repair** is responsible for the integrity of the cellular genome, and its malfunction is a notorious cancer hallmark [9]. **Reproduction** pathway mixes the genomes of two individuals creating a new organism [19]. The mutations in BRCA, that belong to the Reproduction pathway, and their role in fertility is studied by [4]. **Gene expression (Transcription)** governs the transcription and translation that are fundamental cellular processes for protein production of cells and are used as inhibitors in cancer treatment [11]. **Developmental Biology** capture the array of processes by which a fertilized egg gives rise to the diverse tissues of the body. Two processes that are directly involved in this pathway are the regulation of stem cells and activation of HOX genes [21]. These processes are investigated in [2] for their potential in novel treatments for cancer. The malfunction of **Programmed Cell Death** pathway allows the cell to grow uncontrolled and is a cancer hallmark present in most, if not all, types of cancer [6]. This pathway is reported as one of most discussed in cancer therapy [15].

2.3   Super Pathways Sub-networks Enriched with Drivers Information

The driver cancer genes used in this work came from two different databases: The Network of Cancer Genes (NCG) and IntOGen. NCG is a manually curated repository of 2372 genes whose somatic modifications have known or predicted cancer driver roles [22]. From these 2372 genes, 711 are considered known drivers and 1661 candidate drivers. IntOGen defines a pipeline applied to somatic mutations of more than 28000 tumours of 66 cancer types that reveals 568 cancer drivers genes [12]. We combine the 711 genes from NCG and the 568 from IntOGen to found 866 unique drivers used to enrich the Super Pathways networks.

The discovery of cancer driver genes is an open research field. The sets of drivers discussed in this subsection are not exhaustive. From now on, we use the term **Drivers** for all nodes reported as cancer driver genes, and **Non Drivers** for all other nodes, even though there may be undiscovered drivers in this group.

Figure 2 is a visual representation for the Programmed Cell Death Super Pathway Network, that contains 201 nodes and 2636 links. The 27 Drivers in this pathway are marked as red nodes and have a greater size than the 174 blue nodes representing Non Drivers.

3   Results

Regarding the hypothesis of this work which driver genes are topologically different from other genes in the same pathway, we performed topological analyses

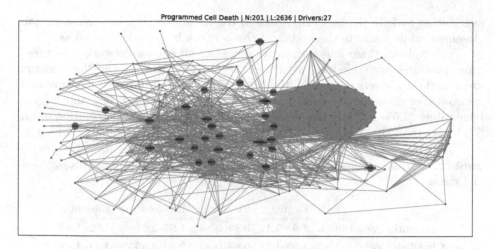

Fig. 2. Programmed cell death network with cancer driver genes enrichment

on the networks presented in Table 1. Such analyses shows a comparison of topo-logical role of Drivers against the Non Drivers in the Network Largest Connect Component (LCC) for each Super Pathway.

Complex networks model the intrinsic relationships of complex systems that can not be fully understood from a single perspective. The Subsect. 3.1 aims to characterize the Driver at the node level using four centrality measures, while Subsect. 3.2 discusses the Drivers' impact on the network resilience using attacks.

3.1   Centrality Measures

A single node can assume multiples roles in the network. Centrality measures capture this roles and are a way to quantity and compare nodes [17]. Since each measure analyzes the nodes by one perspective, it is important to use more than one centrality measure to characterize them. Although there are more than 400 measures [7][4], many of them are for specific types of networks, and there is a high correlation among these measures, especially with the degree [17].

We extract ten centrality measures to characterize the Drivers and Non Drivers: average neighbors degree, betweenness, bridging, closeness, clustering, degree, eccentricity, eigenvector, kcore, and leverage. We choose four classic mea-sures that show a low correlation with each other in Super Pathways networks, indicating that they extract distinct characteristics: **Degree** indicates how many neighbors a node has; **Clustering** define how the neighbors of a node are inter-connected. Group of nodes that cluster together create communities, indicating that they have something in common; **Betweenness** measure bottlenecks by scoring nodes that frequently appear in shortest-paths between all pairs of nodes. Betweenness is frequently used to identify nodes with a high flow of informa-tion; **Closeness** measure the average minimum path length from each node to

---

[4] https://www.centiserver.org/centrality/list/.

every other node in the network. Nodes at the network's periphery have a small closeness, while nodes in the middle of the network have a high closeness.

Table 2 shows these four measures applied to all chosen networks. As these some measures are scale-free, we divide the Drivers mean by the Non Drivers mean and the Drivers median by the Non Drivers median for each measure and network. For example, the table cell from Chromatin Organization and Degree have values "1.09–1.12". In this network, Drivers have a degree distribution mean 9% bigger than the Non Drivers, and a degree distribution median 12% bigger.

Table 2. Centrality measures: drivers mean and median divided by Non Drivers mean and median

	Degree	Clustering	Betweenness	Closeness
Chromatin organization	1.09–1.12	0.89–0.84	1.05–2.57	1.02–1.07
Circadian clock	1.50–1.78	0.89–0.89	2.89–4.86	1.11–1.14
DNA repair	1.09–1.31	0.91–0.95	1.10–0.97	1.02–1.03
Reproduction	1.55–1.56	0.88–0.81	1.47–6.65	1.16–1.15
Gene expression	1.55–2.24	0.84–0.71	2.96–11.5	1.10–1.10
Developmental biology	1.55–1.79	0.76–0.64	3.86–5.83	1.14–1.12
Programmed cell death	1.36–2.07	0.61–0.60	7.74–18.1	1.13–1.12
All super pathways	2.30–3.61	0.73–0.69	5.29–12.3	1.12–1.10

Drivers show a greater degree, especially in All Super Pathways and Gene Expression. Closeness has similar values for the mean and the median. Albeit the percentual difference is subtle, these networks have a small diameter (small world effect), meaning that Drivers are more central. Betweenness is a sensitive measure, prone to outliers. Programmed Cell Death has the most significant values, with the Driver median being 18 times greater.

The prevalence of a higher betweenness in Drivers indicates flow of information. Usually, high betweenness and small clustering are signs of bridges (a single node that connects two "isles"), but this does not fit with the higher degree. A possible interpretation based on Table 2 is that Drivers are hubs (degree) positioned far from the periphery (closeness) that connect different parts (clustering and degree), working as information gateways (betweenness). A visual inspection of Fig. 2 shows that Drivers DCC, LMNA, SATB1, and CDKN2A do not follow this interpretation, but APC, CTNNB1, PAK2, TP53, and BIRC3 does. This interpretation is a generalized idea based on four measures, not a rule.

To further investigate the distinction between Drivers and Non Drivers using centrality measures, we visualize their normalized distributions. The analyses also evidence the scale-free nature of some distributions. Figure 3 and 4 show these analyses for all chosen networks.

The blue boxes represent Non Drivers, and the red boxes the Drivers. Since the set of Drivers are considerably small than Non Drivers, as shown in Table 1, their difference could be random. To address this issue, we also present random

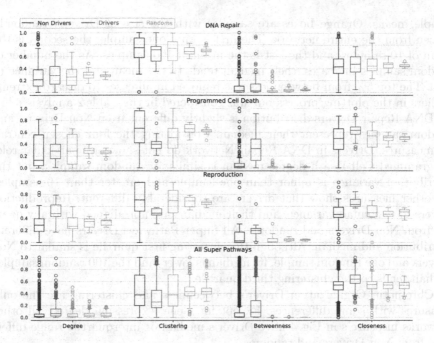

Fig. 3. Centrality measures distribution for programmed cell death, DNA repair, reproduction and all super pathways network.

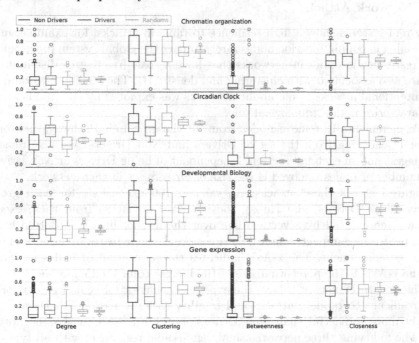

Fig. 4. Centrality measures distribution for chromatin organization, circadian clock, development biology and gene expression.

sample means. Orange boxes are samples with the size of Drivers, randomly chosen from the entire network. The first is only one sample, the second is the mean of 10 samples, and the last is the mean of 100 samples. As the number of random samples increases, their median tends to the mean and the variance to zero. The top 5% from betweenness was removed in all cases to discard extreme outliers in the plotting process but was considered in the Table 2 analysis.

DNA Repair Drivers distribution is slightly different from Non Drivers, and random samples represent their median. Considering the four chosen centrality measures, Drivers in DNA Repair Network do not show any particular role. Programmed Cell Death degree median is similar to random samples, but the distribution variance is minor than one sample and greater than 10 samples. All other measures show that drivers are considerably different. Reproduction drivers have a small variance, and their medians are not aligned with any quartile from Non Drivers and randoms. All Super Pathways Drivers have different distribution and median. Although the clustering first quartile is similar to Non Drivers and one random sample, its median is lower, and the 100 random samples median tends to the clustering third quartile.

Chromatin Organization Driver's betweenness and clustering are the only measures with some differences. Except for Gene Expression degree, all other networks measure's in Fig. 4 have Driver's median or interquartile range difference from Non Drivers and randoms.

3.2 Network Attack

Many real systems show a high resilience to random attacks, for example, malfunctioning in communication nodes rarely impact the global system functioning. This resistance is found in networks that have a scale-free degree distribution, and are also associated with clustering and density [1]. The resilience to random and intentional attacks in metabolic networks was explored by [13] and is a way to characterize nodes' topological role.

We incrementally remove nodes from all chosen Super Pathways networks randomly choosing from the set of Drivers and Non Drivers. We also remove the biggest hub in each interaction, independent of being Drivers or Non Driver. The number of nodes removed is the size of Drivers set in each network.

Figure 5 shows these attacks. As we run 100 executions for each group (Drivers, Non Drivers, and Hubs), the standard deviation from these executions are represented by a vertical line over the impact line. The dotted black line indicates zero impact, removal does not break the network in more than one connected component. The x-Axis indicates the percentual of nodes removed, and the y-Axis is the percentual size of the largest connected component.

The first three networks have the expected behavior for scale-free networks: high resilience to random attacks and fragility to hubs attacks [1]. Drivers are slightly more impactful than Non Drivers, but both are close to the dotted line. The following three networks show remarkable resilience, with all types of attacks being close to zero impact. Development Biology has a global clustering of 53% and 0.04% density. After removing 137 nodes (14%) in each attack, the

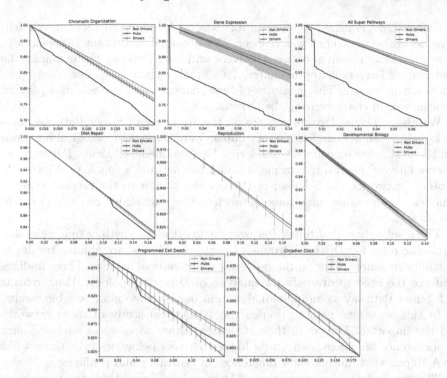

Fig. 5. Random and intentional attacks. The X-axis shows the percentual node removal, and the Y-axis is the percentual size of the largest connect component.

network barely created new connected components. The last two networks show distinct resilience to each type of attack. Drivers removal in Programmed Cell Death impact the network more than Hubs, which is unexpected for scale-free networks. The eighth and ninth removal of hubs are responsible for the firsts breaks. A further investigation shows that these two hubs are the genes *CTNNB1* and *CASP3*, both reported as Drivers, respectively, an oncogene and a tumor suppressor gene.

DNA Repair and Programmed Cell Death are two critical pathways in cancer that show distinct topological behavior. They have the least and most difference in centrality measures when comparing Drivers and Non Drivers. Albeit their respective global clustering, 70% and 61%, and density, 16% and 13%, are similar, the resilience to intentional attacks differ. Reproduction resilience is comparable to DNA Repair, while their Drivers' centrality behavior diverges, being more similar in this aspect to Programmed Cell Death.

4   Discussion and Conclusion

Cancer is a disease with several mechanisms involved that impact its development as a whole. Therefore, it is quite adequate to study it from the perspective of complex systems, making it possible to understand it under different aspects.

Cancer Drivers genes' topological role was analyzed in different Super Pathways networks associated with cancer. We observed significant differences in some centrality measures between Drivers and Non Drivers. For example, the clustering of Drivers is approximately 20% smaller than genes not reported as significant in cancer. The measures of betweenness and closeness also play an essential role in characterizing the Drivers.

We also analyzed Drivers concerning the resilience of Super Pathways networks, which can help understand the impact of mutations in biological functions and their influence on cancer. Considering Programmed Cell Death Pathway, we observe Drivers' central role in maintaining the network's topological integrity. In others networks, the behavior of Drivers was similar to the random removal of nodes. At the same time, some networks show remarkable resilience even to hub attacks.

The results show that Super Pathways networks have distinct topologies and particular roles for Drivers. Groups of pathways that share similar results in centrality measures differ on the resilience of intentional attacks. These findings reinforce the need to diversify the analysis of Driver's topology. Also, treating each Super Pathway as an individual system may provide more reliable results.

In this sense, we plan to deepen the topological study of these networks and the impact of Drivers in their structure. Other topological analyses, such as persistence homology, can unveil hidden Drivers behaviors in pathways like DNA Repair that did not show differences in centrality and resilience.

The topological characterization of Drivers is an important step to comprehend the role that cancer genes play in the cell complex system. The patterns found can be used to identify and validate unknown Drivers that share similar behaviors, like the 1661 candidate drivers from NCG.

Supplementary Information: Source codes, scripts of experiments and the complete list of libraries and versions used in this work are available on the following link: https://github.com/RodrigoHenriqueRamos/BSB-2021-Topological-Characterization-of-Cancer-Driver-Genes.

References

1. Albert, R., Jeong, H., Barabási, A.L.: Error and attack tolerance of complex networks. Nature **406**(6794), 378–382 (2000)
2. Bhatlekar, S., Fields, J.Z., Boman, B.M.: Role of hox genes in stem cell differentiation and cancer. Stem Cells Int. 2018 (2018)
3. Cutigi, J.F., et al.: Combining mutation and gene network data in a machine learning approach for false-positive cancer driver gene discovery. In: Setubal, J.C., Silva, W.M. (eds.) BSB 2020. LNCS, vol. 12558, pp. 81–92. Springer, Cham (2020). https://doi.org/10.1007/978-3-030-65775-8_8
4. Daum, H., Peretz, T., Laufer, N.: BRCA mutations and reproduction. Fertil. Steril. **109**(1), 33–38 (2018)
5. García-Campos, M.A., Espinal-Enríquez, J., Hernández-Lemus, E.: Pathway analysis: state of the art. Front. Physiol. **6**, 383 (2015)
6. Hanahan, D., Weinberg, R.A.: The hallmarks of cancer. Cell **100**(1), 57–70 (2000)

7. Jalili, M., et al.: CentiServer: a comprehensive resource, web-based application and R package for centrality analysis. PLoS ONE **10**(11), e0143111 (2015)
8. Jassal, B., et al.: The reactome pathway knowledgebase. Nucl. Acids Res. **48**(D1), D498–D503 (2020)
9. Jin, M.H., Oh, D.Y.: ATM in DNA repair in cancer. Pharmacol. Ther. **203**, 107391 (2019)
10. Khatri, P., Sirota, M., Butte, A.J.: Ten years of pathway analysis: current approaches and outstanding challenges. PLoS Comput. Biol. **8**(2), e1002375 (2012)
11. Laham-Karam, N., Pinto, G.P., Poso, A., Kokkonen, P.: Transcription and translation inhibitors in cancer treatment. Front. Chem. **8**, 276 (2020)
12. Martinez-Jimenez, F., et al.: A compendium of mutational cancer driver genes. Nat. Rev. Cancer **20**(10), 555–572 (2020)
13. de Mello Pessoa, V.H., Ferreira, C.d.O.L.: Resilience and structure of metabolic networks. Proc. Ser. Braz. Soc. Comput. Appl. Math. **6**(2) (2018)
14. Milenković, T., Memišević, V., Bonato, A., Pržulj, N.: Dominating biological networks. PLoS ONE **6**(8), 1–12 (2011). https://doi.org/10.1371/journal.pone.0023016
15. Mishra, A.P., et al.: Programmed cell death, from a cancer perspective: an overview. Mol. Diagn. Ther. **22**(3), 281–295 (2018)
16. Nussinov, R., Jang, H., Tsai, C.J., Cheng, F.: Precision medicine and driver mutations: computational methods, functional assays and conformational principles for interpreting cancer drivers. PLoS Comput. Biol. **15**(3), e1006658 (2019)
17. Oldham, S., Fulcher, B., Parkes, L., Arnatkevicute, A., Suo, C., Fornito, A.: Consistency and differences between centrality measures across distinct classes of networks. PLoS ONE **14**(7), e0220061 (2019)
18. Ozturk, K., Dow, M., Carlin, D.E., Bejar, R., Carter, H.: The emerging potential for network analysis to inform precision cancer medicine. J. Mol. Biol. **430**(18), 2875–2899 (2018)
19. Reactome: Reproduction (2006). https://reactome.org/content/detail/R-HSA-1474165
20. Reactome: Chromatin organization (2011). https://reactome.org/content/detail/R-HSA-4839726
21. Reactome: Developmental biology (2011), https://reactome.org/content/detail/R-HSA-1266738
22. Repana, D., et al.: The network of cancer genes (NCG): a comprehensive catalogue of known and candidate cancer genes from cancer sequencing screens. Genome Biol. **20**(1), 1–12 (2019)
23. Schuster-Böckler, B., Lehner, B.: Chromatin organization is a major influence on regional mutation rates in human cancer cells. Nature **488**(7412), 504–507 (2012)
24. Shafi, A.A., Knudsen, K.E.: Cancer and the circadian clock. Cancer Res. **79**(15), 3806–3814 (2019)
25. Stratton, M.R., Campbell, P.J., Futreal, P.A.: The cancer genome. Nature **458**(7239), 719–724 (2009)
26. Wu, G., Dawson, E., Duong, A., Haw, R., Stein, L.: Reactomefiviz: a cytoscape app for pathway and network-based data analysis. F1000Research **3** (2014)
27. Wu, G., Feng, X., Stein, L.: A human functional protein interaction network and its application to cancer data analysis. Genome Biol. **11**(5), 1–23 (2010)

Bioinformatics and Computational Biology

CellHeap: A Workflow for Optimizing COVID-19 Single-Cell RNA-Seq Data Processing in the Santos Dumont Supercomputer

Vanessa S. Silva[1], Maiana O. C. Costa[2], Maria Clicia S. Castro[4],
Helena S. Silva[3], Maria Emilia M. T. Walter[3], Alba C. M. A. Melo[3],
Kary A. C. Ocaña[2], Marcelo T. dos Santos[2], Marisa F. Nicolas[2],
Anna Cristina C. Carvalho[1], Andrea Henriques-Pons[1],
and Fabrício A. B. Silva[1(⊠)]

[1] Oswaldo Cruz Foundation, Rio de Janeiro, Brazil
fabricio.silva@fiocruz.br
[2] National Laboratory for Scientific Computing, Petropolis, Brazil
[3] University of Brasilia, Brasilia, Brazil
[4] State University of Rio de Janeiro, Rio de Janeiro, Brazil

Abstract. Currently, several hundreds of Terabytes of COVID-19 single-cell RNA-seq (scRNA-seq) data are available in public repositories. This data refers to multiple tissues, comorbidities, and conditions. We expect this trend to continue, and it is realistic to predict amounts of COVID-19 scRNA-seq data increasing to several Petabytes in the coming years. However, thoughtful analysis of this data requires large-scale computing infrastructures, and software systems optimized for such platforms to generate biological knowledge. This paper presents CellHeap, a portable and robust workflow for scRNA-seq customizable analyses, with quality control throughout the execution steps and deployable on supercomputers. Furthermore, we present the deployment of CellHeap in the Santos Dumont supercomputer for analyzing COVID-19 scRNA-seq datasets, and discuss a case study that processed dozens of Terabytes of COVID-19 scRNA-seq raw data.

Keywords: Single-cell RNA-seq · Bioinformatics workflow ·
COVID-19 · High-performance computing

1 Introduction

Gene expression is a highly heterogeneous biological process, even among similar cell types. An accurate comprehension of the transcriptome of individual cells is essential to elucidate its role in biological functions and to understand how gene expression can promote beneficial or harmful states at the tissue and/or organism level. While conventional bulk RNA sequencing (RNA-seq) can only provide the average expression signal for a set of cells, single-cell RNA sequencing

© Springer Nature Switzerland AG 2021
P. F. Stadler et al. (Eds.): BSB 2021, LNBI 13063, pp. 41–52, 2021.
https://doi.org/10.1007/978-3-030-91814-9_4

(scRNA-seq) describes the state of individual cells with extraordinary resolution. Currently, the outcomes of this analysis can support various studies from microbial population cells (e.g. bacteria, trypanosomatids, unicellular alga, yeast) to mammalian tissues, making it possible to identify the high heterogeneity of cell populations in different conditions [12,15,20,31].

Briefly, scRNA-seq sequencing consists of several steps, mainly: i) isolating single cells, ii) cell lysis, iii) mRNA hybridization, iv) reverse transcription, v) PCR amplification, and vi) sequencing and analysis, where each mRNA is mapped to its cell-of-origin and gene-of-origin, and each cell's pool of mRNA can be analyzed. Numerous sequencing protocols have been improved in recent years, allowing us to develop a better knowledge of biological systems of the cell [15,31]. A pleitora of scRNA-seq tools have been converged in the development of specific scientific workflows applied in a variety of studies, including COVID-19. The tools and protocols in different programming languages are quite a massive field. The user should check which tool is more suitable for answering questions from his/her study, and which language is most appropriate [31].

Scientific workflows (or pipelines) are composed of a set of interconnected and automated tasks, which are run according to their input/output dependencies [1]. They are a powerful and successful apparatus, executed several times on daily to solve problems in various research domains. To create a workflow, scientists often use an *ad hoc* strategy, writing from scratch a set of scripts that are used to connect the tasks. This approach leads to many workflows, each one targeting a specific situation, usually leading to a myriad of slightly different variations of the same idea. As a consequence, productivity is reduced, and the production of results may take much longer than needed [5].

Nevertheless, most of the workflows used in the COVID-19 studies are static, built in an *ad hoc* manner, and run in a standalone desktop machine [25–27,34]. In a scenario where sophisticated bioinformatics tools are created regularly, we claim that scientific workflows must be flexible, enabling the user to choose among different tools that execute the same task. In addition, workflows should be extensible, allowing new modules or phases to be integrated into them. Moreover, we argue that a workflow for scRNA-seq studies that analyses samples of hundreds of Gigabytes should be executed in a high-performance computing platform, such as a supercomputer.

Here, we present CellHeap, a portable and robust workflow for scRNA-seq customizable analysis, with quality control throughout the execution steps, which ensures reliable results, and runs on supercomputers. Our workflow is applicable to scRNA-seq derived from droplet-based methods with UMIs, such as those related to 3'-end (e.g. Drop-seq [21]) and 5'-end (e.g. STRT-seq [13]) transcript sequencing technologies. CellHeap starts by performing the scRNA-seq samples' download and finishes with clustering processes and some complex biological analysis. Our workflow integrates R tools, other bioinformatics libraries and software, and is customizable depending on the purpose of each analysis.

2   Description of the CellHeap Workflow

This section presents CellHeap, a flexible, extensible, portable, and robust work-flow for scRNA-Seq customized bioinformatics analysis. CellHeap allows quality control to ensure reliable results and is focused on high capability computational support for parallelizing and distributing workflow tasks.

CellHeap is composed of five phases (shown in Fig. 1): sample curation; gene count matrix generation; quality control; dimensionality reduction and cluster-ing analysis; and advanced cell level and gene level analysis. In addition to these phases, we also provide an optional samples aggregation phase. Its code is pro-vided at https://github.com/FioSysBio/CellHeap.

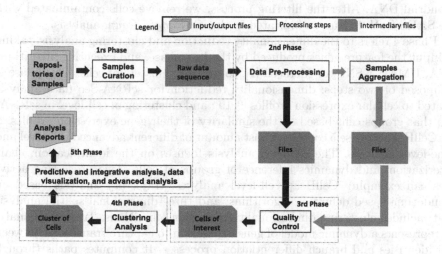

Fig. 1. CellHeap workflow conceptual view. Dashed rectangles identify the different phases of the workflow.

Phase 1 refers to samples curation processes of the scRNA-Seq input datasets related to organisms and tissues of interest. It follows the inclusion/exclusion cri-teria established for the Expression Atlas [24] of the European Bioinformatics Institute (EMBL-EBI) [14]. At the end of this phase, raw data sequences are selected according to the quality criteria of samples. In this project, the adopted main criteria are: (i) to link to supplementary files on the measurement of genes of experiments; (ii) to access the sample's raw data through SRA selector links; (iii) to verify if all series samples belong to single species; (iv) to verify whether samples come from non-bacterial species; (v) to verify the source of the descrip-tion of experiments, protocols, cell line information, cell type, and disease, as listed by Experimental Factor Ontology (EFO) [22] or publication data; (vi) to check whether metadata matches the samples' names; (vii) to verify whether scRNA-Seq experiments use protocols of Smart-seq2, Smart-like, Drop-seq, Seq-well, $10 \times$ V2 (3 and 5 prime), or $10 \times$ V3 (3 prime).

Phase 2 aims to generate the gene count matrix, which can be executed by CellRanger[1]. *CellRanger Count* requires a reference genome and executes several quality checks on both the dataset and the reference genome.

Phase 3 is related to the quality control of the single-cell data. Its input is the gene-count matrix produced by phase 2, and it executes the package Seurat [29] to produce filtered infected/uninfected cells. Filtering parameters remove low-quality cells and doublets (or multiplets) which are artifacts from two (or more cells) erroneously treated as a single cell with high gene counts. The filtering parameters are adjustable to control the cell quality according to the project's objectives. We filtered cell matrices removing cell barcodes according to the three criteria (covariates) commonly used for scRNA-Seq quality control processing [19]: UMI counts; genes expressed per cell; and percentage of mitochondrial DNA. After the filtering process, we remove cells contaminated with the SARS-CoV-2 virus, as indicated in the *Cellranger Count* analysis.

Phase 4 refers to the dimensionality reduction and clustering analysis, taking as input the feature genes produced by the last phase, and executing algorithms PCA, UMAP, t-SNE, or Metacell [2], to produce clusters of cells. This phase is composed of two steps: dimensionality reduction for scRNA-Seq data analysis related to cellular expression profiles [9,19]; and clustering of cell-level in scRNA-Seq that groups cells based on the similarity of their gene expression profiles.

CellHeap's phase 5 contains a vast amount of different advanced cell-level and gene-level analyses. The cell-level analysis focuses on the identification, characterization, and dynamics inference of groups of cells. Two processing activities can exemplify CellHeap cell-level analyses: trajectory analysis based on pseudotime-based definition algorithms; and three-dimensional spatial analysis that includes clustering dimensionality-reduction algorithms. Trajectory analysis represents a dynamic model of gene expression to capture transitions between cell identities and branch differentiation processes. It computes paths through cellular space that minimizes transcriptional changes among neighboring cells. A pseudotime variable describes the evolution of transcriptional states along the trajectory under certain conditions of developmental time [10,28,33]. CellHeap may execute three-dimensional spatial analysis for single cells, which explores specific characteristics of a cell or a set of cells, such as cell-cell interactions.

The gene-level analysis focuses on analyzing the expression variation of sets of genes among groups of cells. The CellHeap uses three sets of algorithms for this analysis: differentially expressed genes (DEG) analysis; gene set enrichment analysis; and network inference of interacting genes. In scRNA-Seq processing, DEG analysis studies variations of gene expression between two conditions. The

---

[1] Cellranger is a set of analysis pipelines that process Chromium single-cell data to align reads, generate feature-barcode matrices, perform clustering and other secondary analysis, and more. *CellRanger Count* is executed once for each dataset, and *CellRanger Aggregate* is optionally executed for aggregating several different datasets/tissues. In addition, *CellRanger Count* and *CellRanger Aggregate* generate a gene-count matrix, where the results depend on the analysis performed in a simple or an aggregated way.

expression variation is counted in clusters of single cells, representing different cell types or experimental conditions. As a general recommendation, differential expression requires non-corrected inputs matrices for batch effects or input matrices that incorporate technical covariates [19]. Gene set enrichment analysis aims to group DEGs as they are involved in common biological processes. The CellHeap workflow can use available resources as Gene Ontology (GO), Kyoto Encyclopedia of Genes and Genomes (KEGG), Molecular Signatures Database (MSigDB) [18], and Reactome [6]. Networks of interacting genes (pathways and DEGs location) analysis support identifying and visualizing the main pathways involved, and the DEGs' localization in these pathways. The CellHeap can use available resources as PANTHER [23], Database for Annotation, Visualization and Integrated Discovery (DAVID) [11], The Reactome pathway analysis [6] and ReactomeGSA [32].

3 Results

This section describes a practical experiment with the CellHeap workflow - a COVID-19 single-cell RNA-seq Big Data scenario on the Santos Dumont Supercomputer. Our group had identified several hundreds of Terabytes of COVID-19 scRNA-seq data available in public repositories up to this date. This data relates to multiple tissues, comorbidities, and conditions. The thoughtful analysis of this data to generate biological knowledge requires large-scale computing infrastructures, and software systems optimized for platforms such as CellHeap.

3.1 Input Data and Experiment Setup

We took as input the bronchoalveolar scRNA-seq dataset GSE145926 from the NIH GEO database, used by Liao et al. [17]. This dataset encompasses tens of terabytes of raw data.

The experiment focus on the use of the CellHeap workflow in a scRNA-Seq analysis from droplet-based methods with UMIs, such as those related to 3′-end (e.g., Drop-seq [21]) and 5′-end (e.g., STRT-seq [13]) transcript sequencing technologies. In what follows, we detail all the adopted tools, as guided by the CellHeap workflow.

Phase 1. We selected the cataloged series of the Gene Expression Omnibus (GEO) repository [4]. The results presented in Liao et al. [17] considered 13 patients, where 3 were controls, 3 presented mild symptoms, and 6 six patients developed severe symptoms of COVID-19. However, when we applied the dataset selection criteria (Sect. 2), one control dataset was discarded since the raw data for the control dataset ID GSM3660650 was not available. Therefore, in the following analysis, different from Liao et al. [17], we considered only 12 patients.

Phase 2. We downloaded the NCBI Sequence Read Archive (SRA) raw data in FASTQ format files, with the NCBI SRA toolkit fastq-dump tool [30]. Each downloaded sample had two associated read files, one containing information about the UMI while the second file contained the transcription sequences.

We used Cellranger v4.0.0 from 10X Genomics [35] to process scRNA-Seq data, and generate the gene-barcode matrix. Cellranger requires high computational power, demanding a large amount of storage and high information processing capacity. Some observations about the Cellranger scalability follow. Cellranger supports a Job Submission Mode and a Cluster-Mode. Both running modes are adequate for deploying Cellranger in clusters. These modes allow Cellranger to efficiently use the computing power available in clusters, also representing a significant advantage in big data scenarios, and over competing tools. This phase is the most intensive computing among all the CellHeap's phases. The efficient use of the available computing power is a fundamental requirement, as described in the big data deployment scenario. Besides, Cellranger requires access to the raw data, demanding a large storage capacity. Therefore, the deployment of this activity in the Santos Dumont supercomputer was mandatory.

Each dataset analyzed by Cellranger generates quality assessment reports that display important quality metrics. For instance, Q30 measures the fraction of bases with a Q-score of at least 30 in the cell barcode, RNA sequences and sample index sequences. All the output reports relative to the results of this section are available at the repository. We built a hybrid reference that includes both the human genome (GRCh38 v3.0.0) and the SARS-CoV-2 genome (NC_045512v2) to generate the results. Cellranger's output report also informs how many cells contain genetic material from the virus.

Phase 3. We filtered the cell matrices by removing cell barcodes according to the criteria already described in the CellHeap framework: (i) UMI counts; (ii) genes expressed per cell; and (iii) percentage of mitochondrial DNA. These three covariates are commonly used for scRNA-Seq quality control processing [19].

We used Seurat v4.0 [8] for single-cell quality control (QC). The output matrices produced from Cellranger Count (or Cellranger Aggregate) are inputs to the Seurat package. After this filtering process, considering the expression profile corrected, we created a Seurat object using the CreateSeuratObject function. The output of this step generated the gene count matrix.

After QC, we filtered cells contaminated with SARS-CoV-2, and generated two files for further processing: one containing only non-infected cells; and the other one only infected cell data.

Phase 4. For the case study described in this paper, we used MetaCell [2] in the clustering phase. The Metacell algorithm aims to cluster cells with homogeneous transcriptional profiles into groups, called metacells. The overall idea is that cells belonging to a metacell could have been resampled from the same cell. Metacell requires a feature gene set as additional input. Metacell processing results depend on feature genes selection, which may be defined by the user, related to the biological question of the analysis, or generated automatically, based on gene markers with high variance. For this case study, we chose Metacell clustering parameters other than feature genes according to Bost et al. [3]. In addition, Metacell also applies an outlier filtering step to identify outlier cells and doublets, then filtering the resulting clusters.

Phase 5. The CellHeap workflow allows a variety of advanced cell-level and gene-level analyses. In this paper, we illustrate phase 5 with several extended analyses from Metacell. For instance, Metacell can generate heatmaps of genes and metacells. Another type of analysis is the 2D projections of cells and metacells, resulting in graphs where edges indicate metacells with similar transcriptional profiles. In addition, Metacell generates log2 fold enrichment bar plots for a specific gene over the median expression values of all the metacells in the same analysis. Finally, another analysis is the hierarchical clustering of metacells used in this paper to classify and color metacells according to cell type.

Beyond Metacell, we also used in this paper an automatics procedure to identify cell types based on queries to the Enrichr [16] web application and the PanglaoDB [7] repository.

3.2 Environmental Setup

We deployed the CellHeap workflow in the Santos Dumont (SDumont) Supercomputer[2]. SDumont has an installed processing capacity in the order of 5.1 Petaflop/s, presenting a hybrid configuration of computational nodes regarding the available parallel processing architecture. It has 36,472 CPU cores, distributed across 1,134 computational nodes, of which the majority are composed exclusively of CPUs with a multi-core architecture. In addition, the SDumont Supercomputer has a Lustre parallel file system, integrated with the Infiniband network, with a raw storage capacity of 1.7 PBytes and a secondary file system with a raw capacity of 640 TBytes.

Our group has Premium access to SDumont, which provides allocations to research projects that request more than 5,000,000 Allocation Units (AU). Each AU corresponds to the use of one core of a B710 compute node for one hour. The B710 compute nodes feature 2 Intel Xeon E5-2695v2 Ivy Bridge CPUs (12 cores @2.4 GHz) and 64 Gb RAM. The running mode of Cellranger in the SDumont supercomputer is the Job Submission mode. The workload manager available in the SDumont supercomputer is Slurm[3]. At this moment, Cellranger's Cluster-Mode does not support Slurm.

3.3 Results Discussion

Phase 1 took up 48 h to execute for each of the 12 raw data samples analyzed in this paper, on a B170 compute node. For Phase 2, *Cellranger Count* required over 12 h of computing processing (not considering queuing time), on a B170 compute node, in the worst case for the datasets processed in this paper.

For Phase 3 processing, we mainly used Seurat functions. Regarding execution time, both Phase 3 and Phase 4/5 (Metacell) scripts execute much faster when compared to execution times of Phases 1 and 2. Typically, Phase 3 and Phase 4 scripts demanded less than 10 min on a B170 node.

---

[2] Supercomputer details in https://sdumont.lncc.br.
[3] Slurm details in https://slurm.schedmd.com.

Regarding metacell processing, we used *mcell_gset_filter_varmean* and *mcell_gset_filter_cov* functions for automatically generating the gene features, used as input for metacells definition. We set the threshold on the variance/mean ratio to 0.4 for this BALF dataset, as in [3]. For instance, this process resulted in a feature set composed of 135 genes for the dataset of patients with mild symptoms.

After building the K-nn cell similarity graph, resampling graph partitions, and filtering parametric outliers from the metacell cover, we performed hierarchical clustering on the final set of metacells, to perform a systematic annotation procedure. The results of this process are a "confusion matrix" and the corresponding cluster hierarchy, as shown in Fig. 2. The confusion matrix is a metacell pairwise similarity matrix, which summarizes the K-nn graph connectivity among all cells in each pair of metacells. Metacells are then hierarchically clustered based on this confusion matrix. In Fig. 2, on the left of the cluster hierarchy (top of the figure), the top 5 marker genes for each cluster are shown. The markers on the left are the genes that maximize the average log-fold enrichment of the metacells composing the cluster over all other metacells. Other markers (e.g., related to sibling subtrees, in gray) are shown on the right side. For systematic annotation of metacells, we use the markers on the left side.

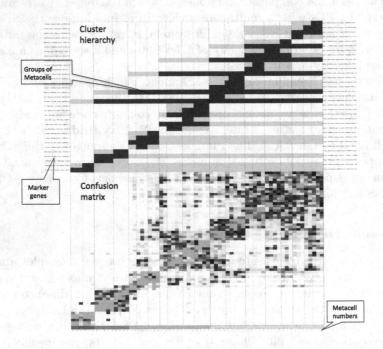

Fig. 2. Hierarchical Clustering (top) and confusion matrix (bottom) for the dataset corresponding to patients with mild symptoms. Each blue bar on the top represents a cluster of metacells. The normalized edge count is minimal (0) for white dots and maximal (1) for yellow dots in the confusion matrix. (Color figure online)

In Fig. 2, there are 65 metacells and 33 clusters of metacells. The largest cluster has 59 metacells, and the smallest one, three. As expected in a hierarchical clustering procedure, several clusters of metacells are proper subsets of larger clusters. The user can adjust the number of clusters through the T_gap parameter in the *mcell_mc_hierarchy*. T_gap defines the minimal branch length for defining the structure of clusters of metacells. For generating Fig. 2, we used the standard value of 0.04 for T_gap.

For each metacell cluster (or *supermetacell*), the hierarchical clustering processing generates sets of gene markers for each cluster (see Fig. 2). These gene markers can be used to classify clusters of metacells, either by a human expert or using automatics tools.

To illustrate this process, we submitted the 20 markers that maximize average log-fold enrichment of the metacells composing the cluster over all other metacells to an automatic annotation website. Enrichr [16] is an interactive gene list enrichment tool. One of the several functionalities of Enrich is to indicate more probable cell types associated with a set of markers, based on queries to several repositories. For this illustrative example, we chose results obtained from PanglaoDB [7] (2021 edition), a repository for exploring human and mouse scRNA-seq data. We did a preliminary annotation of the metacell clusters shown in Fig. 2 using EnrichR/PanglaoDB, and plotted the resulting set of colored metacells in a 2D projection using Metacell's *mcell_mc2d_force_knn* and *mcell_mc2d_plot* functions (see Fig. 3).

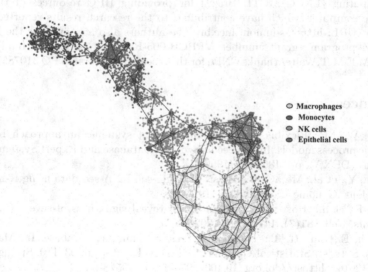

○ Macrophages
● Monocytes
◉ NK cells
● Epithelial cells

Fig. 3. 2D Projection of cells and metacells - COVID-19 Patients with mild symptoms. This projection includes only cells not infected with SARS-CoV-2. Colored dots are cells, while circles represent metacells. Automated cell type classification and coloring were based on queries to Enrichr/PanglaoDB websites. (Color figure online)

As is the case with most automated methods, algorithmic-generated classification results should be curated by human experts. Refinements are possible in the classification presented in Fig. 3. In Fig. 2, it is possible to identify proper subsets of the clusters defined in Fig. 3, which may represent opportunities for more specific classifications. Nevertheless, since one of the objectives of Cell-Heap is to maximize automation in scRNA-seq processing, Fig. 3 illustrates the possibility of automating the analysis up to cell type annotation.

4 Conclusion

This paper presented CellHeap, a portable, flexible workflow for scRNA-seq data processing deployable in supercomputers. Bioinformatics tools for scRNA-seq are a very active area of research today, and the number of new tools has increased substantially in the last few years. On the other hand, the amount of scRNA-seq data available for analysis in public repositories has also been increasing quickly. Therefore, flexible, robust, and scalable workflows are paramount to take advantage of this massive amount of data to increase biological knowledge.

We expect CellHeap to improve continuously, aggregating new tools as they are validated and become available. We also expect to deploy CellHeap in other high-throughput platforms, such as computing clouds, soon.

Acknowledgments. The authors acknowledge the National Laboratory for Scientific Computing (LNCC/MCTI, Brazil) for providing HPC resources of the SDumont supercomputer, which have contributed to the research results reported within this paper. URL: http://sdumont.lncc.br. The authors also acknowledge the INOVA-FIOCRUZ program (grant number VPPCB-005-FIO-20-2-34-52) for funding this research. M.E.M.T.Walter thanks CNPq for the research scholarship PQ 310785/2018-9.

References

1. Aalst, W.M.P.: Flexible workflow management systems: an approach based on generic process models. In: Proceedings of the Database and Expert Systems Applications (DEXA), pp. 186–195 (1999)
2. Baran, Y., et al.: MetaCell: analysis of single-cell RNA-seq data using K-nn graph partitions. Genome Biol. **20**(1), 1–19 (2019)
3. Bost, P., et al.: Host-viral infection maps reveal signatures of severe COVID-19 patients. Cell **181**(7), 1475–1488 (2020)
4. Clough, E., Barrett, T.: The gene expression omnibus database. In: Mathé, E., Davis, S. (eds.) Statistical Genomics. MMB, vol. 1418, pp. 93–110. Springer, New York (2016). https://doi.org/10.1007/978-1-4939-3578-9_5
5. Deelman, E., Peterka, T., Altintas, I., et al.: The future of scientific workflows. Int. J. High Perform. Comput. Appl. **32**(1), 159–175 (2018)
6. Fabregat, A., Jupe, S., Matthews, L., Sidiropoulos, K., et al.: The reactome pathway knowledgebase. Nucleic Acids Res. **4**(46(D1)), D649–D655 (2018)
7. Franzén, O., Gan, L.M., Björkegren, J.L.: PanglaoDB: a web server for exploration of mouse and human single-cell RNA sequencing data. Database 2019 (2019)

8. Hao, Y., et al.: Integrated analysis of multimodal single-cell data. Cell (2021)
9. Heimberg, G., Bhatnagar, R., El-Samad, H., Thomson, M.: Dimensionality in gene expression data enables the accurate extraction of transcriptional programs from shallow sequencing. Cell Syst. **2**(4), 239–250 (2016)
10. Herring, C.A., Banerjee, A., McKinley, E.T., et al.: Unsupervised trajectory analysis of single-cell RNA-seq and imaging data reveals alternative tuft cell origins in the gut. Cell Syst. **6**(1), 37–51 (2018)
11. Huang, D., Sherman, B., Lempicki, R.: Systematic and integrative analysis of large gene lists using DAVID bioinformatics resources. Nat. Protoc. **4**, 44–57 (2009)
12. Hwang, B., Lee, J., Bang, D.: Single-cell RNA sequencing technologies and bioinformatics pipelines. Exp. Mol. Med. **50**, 1–14 (2018)
13. Islam, S., et al.: Highly multiplexed and strand-specific single-cell RNA 5′ end sequencing. Nat. Protoc. **7**(5), 813–828 (2012)
14. Kanz, C., Aldebert, P., Althorpe, N., et al.: The EMBL nucleotide sequence database. Nucleic Acids Res. **33**(Suppl_1), D29–D33 (2005)
15. Kuchina, A., et al.: Microbial single-cell RNA sequencing by split-pool barcoding. Science (2020)
16. Kuleshov, M.V., et al.: Enrichr: a comprehensive gene set enrichment analysis web server 2016 update. Nucleic Acids Res. **44**(W1), W90–W97 (2016)
17. Liao, M., et al.: Single-cell landscape of bronchoalveolar immune cells in patients with COVID-19. Nat. Med. **26**(6), 842–844 (2020)
18. Liberzon, A., Subramanian, A., Pinchback, R., Thorvaldsdóttir, H., Tamayo, P., Mesirov, J.P.: Molecular signatures database (MSigDB) 3.0. Bioinformatics **27**(12), 1739–1740 (2011)
19. Luecken, M.D., Theis, F.J.: Current best practices in single-cell RNA-seq analysis: a tutorial. Mol. Syst. Biol. **15**(e8746), 1–23 (2019)
20. Ma, F., Salome, P.A., Merchant, S.S., Pellegrini, M.: Single-cell RNA sequencing of batch chlamydomonas cultures reveals heterogeneity in their diurnal cycle phase. Plant Cell **33**(4), 1042–1057 (2021)
21. Macosko, E.Z., et al.: Highly parallel genome-wide expression profiling of individual cells using nanoliter droplets. Cell **161**(5), 1202–1214 (2015)
22. Malone, J., et al.: Modeling sample variables with an experimental factor ontology. Bioinformatics **26**(8), 1112–1118 (2010)
23. Mi, H., Ebert, D., Muruganujan, A., et al.: PANTHER version 16: a revised family classification, tree-based classification tool, enhancer regions and extensive API. Nucleic Acids Res. **49**(D1), D394–D403 (2020)
24. Papatheodorou, I., Moreno, P., Manning, J., Fuentes, et al.: Expression atlas update: from tissues to single cells. Nucleic Acids Res. **48**(D1), D77–D83 (2019)
25. Schulte-Schrepping, J., Reusch, N., Paclik, D., et al.: Severe COVID-19 is marked by a dysregulated myeloid cell compartment. Cell **182**(6), 1419–1440 (2020)
26. Silvin, A., Chapuis, N., Dunsmore, G., et al.: Elevated calprotectin and abnormal myeloid cell subsets discriminate severe from mild COVID-19. Cell **182**(6) (2020)
27. Song, E., Bartley, C.M., Chow, R.D.: Divergent and self-reactive immune responses in the CNS of COVID-19 patients with neurological symptoms. Cell Rep. Med. **2**(5) (2021)
28. Street, K., Risso, D., Fletcher, R., et al.: Slingshot: cell lineage and pseudotime inference for single-cell transcriptomics. BMC Genomics **19**(477), 1–16 (2018)
29. Stuart, T., et al.: Comprehensive integration of single-cell data. Cell **177**(7), 1888–1902 (2019)
30. SRA Toolkit Development Team: Sra toolkit. http://ncbi.github.io/sra-tools/. Accessed Aug 2021

31. Vigneron, A., et al.: Single-cell RNA sequencing of trypanosoma brucei from tsetse salivary glands unveils metacyclogenesis and identifies potential transmission blocking antigens. Proc. Natl. Acad. Sci. **117**(5), 2613–2621 (2020)

32. Viteri, J.G.G., Sidiropoulos, K., et al.: ReactomeGSA - efficient multi-omics comparative pathway analysis. Mol. Cell. Proteomics **19**(12), 2115–2125 (2020)

33. Wolf, F.A., Hamey, F.K., Plass, M., et al.: PAGA: graph abstraction reconciles clustering with trajectory inference through a topology preserving map of single cells. Genome Biol. **20**(59), 1–9 (2019)

34. Yao, C., Bora, S.A., Parimon, T., et al.: Cell-type-specific immune dysregulation in severely ill COVID-19 patients. Cell Rep. **34**(1) (2020)

35. Zheng, G.X., et al.: Massively parallel digital transcriptional profiling of single cells. Nat. Commun. **8**(1), 1–12 (2017)

Combining Orthology and Xenology Data in a Common Phylogenetic Tree

Marc Hellmuth[1]📧, Mira Michel[2], Nikolai N. Nøjgaard[3]📧, David Schaller[4,5]📧,
and Peter F. Stadler[4,5,6,7,8(✉)]📧

[1] Department of Mathematics, Faculty of Science, Stockholm University,
10691 Stockholm, Sweden
marc.hellmuth@math.su.se

[2] Faculty of Mathematics and Computer Science, Fernuniversität Hagen,
Universitätsstrasse 47, 58097 Hagen, Germany
mira.michel@studium.fernuni-hagen.de

[3] Department of Mathematics and Computer Science, University of Southern
Denmark, Odense M, Denmark

[4] Max Planck Institute for Mathematics in the Sciences, Leipzig, Germany
sdavid@bioinf.uni-leipzig.de

[5] Bioinformatics Group, Department of Computer Science, and Interdisciplinary
Center for Bioinformatics, Universität Leipzig, Härtelstrasse 16-18,
04107 Leipzig, Germany
studla@bioinf.uni-leipzig.de

[6] Institute for Theoretical Chemistry, University of Vienna, Vienna, Austria

[7] Facultad de Ciencias, Universidad Nacional de Colombia, Bogotá, Colombia

[8] Santa Fe Institute, Santa Fe, NM, USA

Abstract. In mathematical phylogenetics, types of events in a gene tree T are formalized by vertex labels $t(v)$ and set-valued edge labels $\lambda(e)$. The orthology and paralogy relations between genes are a special case of a map δ on the pairs of leaves of T defined by $\delta(x,y) = q$ if the last common ancestor $\mathrm{lca}(x,y)$ of x and y is labeled by an event type q, e.g., speciation or duplication. Similarly, a map ε with $m \in \varepsilon(x,y)$ if $m \in \lambda(e)$ for at least one edge e along the path from $\mathrm{lca}(x,y)$ to y generalizes xenology, i.e., horizontal gene transfer. We show that a pair of maps (δ, ε) derives from a tree (T, t, λ) in this manner if and only if there exists a common refinement of the (unique) least-resolved vertex labeled tree (T_δ, t_δ) that explains δ and the (unique) least-resolved edge labeled tree $(T_\varepsilon, \lambda_\varepsilon)$ that explains ε (provided both trees exist). This result remains true if certain combinations of labels at incident vertices and edges are forbidden.

Keywords: Mathematical phylogenetics · Rooted trees · Binary relations · Symbolic ultrametric · Fitch map · Consistency

1 Introduction

An important task in evolutionary biology and genome research is to disentangle the mutual relationships of related genes. The evolution of a gene family

© Springer Nature Switzerland AG 2021
P. F. Stadler et al. (Eds.): BSB 2021, LNBI 13063, pp. 53–64, 2021.
https://doi.org/10.1007/978-3-030-91814-9_5

can be understood as a tree T whose leaves are genes and whose inner vertices correspond to evolutionary events, in particular speciations (where genomes are propagated into different lineages that henceforth evolve independently), duplications (of genes within the same genome), and horizontal gene transfer (where copies of an individual's genes are transferred into an unrelated species) [4]. Mathematically, these concepts are described in terms of rooted trees T with vertex labels t representing event types and edge labels λ distinguishing vertical and horizontal inheritance. On the other hand, orthology (descent from a speciation) or xenology (if the common history involves horizontal transfer events) can be regarded as binary relations on the set L of genes. Given the orthology or xenology relations, one then asks whether there exists a vertex or edge labeled tree T with leaf set L that "explains" the relations [5,9]. Here, we ask when such relational orthology and xenology data are consistent, i.e., when they can be explained by a common tree. A conceptually similar question is addressed in a very different formal setting in [15].

Instead of considering a single binary orthology or xenology relation, we consider here multiple relations of each type. This is more conveniently formalized in terms of maps that assign finite sets of labels. Two types of maps are of interest: Symbolic ultrametrics, i.e., symmetric maps determined by a label at the last common ancestor of two genes [2], generalize orthology; Fitch maps, i.e., non-symmetric maps determined by the union of labels along the path connecting two genes [12], form a generalization of xenology. For both types of maps unique least-resolved trees (minimal under edge-contraction) exist and can be constructed by polynomial time algorithms [2,12]. Here we consider the problem of finding trees that are simultaneously edge- and vertex-labeled and simultaneously explain both types of maps. We derive a simple condition for the existence of explaining trees and show that there is again a unique least-resolved tree among them. We then consider a restricted version of problem motivated by concepts of observability introduced in [17].

2   Preliminaries

Trees and Hierarchies. Let T be a rooted tree with vertex set $V(T)$, leaf set $L = L(T) \subseteq V(T)$, set of inner vertices $V^0(T) := V(T) \setminus L(T)$, root $\rho \in V^0(T)$, and edge set $E(T)$. An edge $e = \{u, v\} \in E(T)$ is an *inner* edge if $u, v \in V^0(T)$. The ancestor partial order on $V(T)$ is defined by $x \preceq_T y$ whenever y lies along the unique path connecting x and the root. We write $x \prec_T y$ if $x \preceq_T y$ and $x \neq y$. For $v \in V(T)$, we set $\mathrm{child}_T(v) := \{u \mid \{v, u\} \in E(T), u \prec_T v\}$ and $\mathrm{parent}_T(u) := v$ for all $u \in \mathrm{child}_T(v)$. All trees T considered here are *phylogenetic*, i.e., they satisfy $|\mathrm{child}_T(v)| \geq 2$ for all $v \in V^0(T)$. The *last common ancestor* of a vertex set $W \subseteq V(T)$ is the unique $\preceq_T$-minimal vertex $\mathrm{lca}_T(W) \in V(T)$ satisfying $w \preceq_T \mathrm{lca}_T(W)$ for all $w \in W$. For brevity, we write $\mathrm{lca}_T(x, y) := \mathrm{lca}_T(\{x, y\})$. Furthermore, we will sometimes write $vu \in E(T)$ as a shorthand for "$\{u, v\} \in E(T)$ with $u \prec_T v$." We denote by $T(u)$ the subtree of T rooted in u and write $L(T(u))$ for its leaf set.

Furthermore, $L_v^T := \{(x, y) \mid x, y \in L(T), \mathrm{lca}_T(x, y) = v\}$ denotes the set of pairs of leaves that have v as their last common ancestor. By construction, $L_v^T \cap L_{v'}^T = \emptyset$ if $v \neq v'$. Since T is phylogenetic, we have $L_v^T \neq \emptyset$ for all $v \in V^0(T)$, i.e., $\mathcal{L}(T) := \{L_v^T \mid v \in V^0(T)\}$ is a partition of the set of distinct pairs of vertices.

A hierarchy on L is set system $\mathcal{H} \subseteq 2^L$ such that (i) $L \in \mathcal{H}$, (ii) $A \cap B \in \{A, B, \emptyset\}$ for all $A, B \in \mathcal{H}$, and (iii) $\{x\} \in \mathcal{H}$ for all $x \in L$. There is a well-known bijection between rooted phylogenetic trees T with leaf set L and hierarchies on L, see e.g. [19, Thm. 3.5.2]. It is given by $\mathcal{H}(T) := \{L(T(u)) \mid u \in V(T)\}$; conversely, the tree $T_{\mathcal{H}}$ corresponding to a hierarchy $\mathcal{H}$ is the Hasse diagram w.r.t. set inclusion. Thus, if $v = \mathrm{lca}_T(A)$ for some $A \subseteq L(T)$, then $L(T(v))$ is the inclusion-minimal cluster in $\mathcal{H}(T)$ that contains A [11].

Let T and T^* be phylogenetic trees with $L(T) = L(T^*)$. We say that T^* is a *refinement* of T if T can be obtained from T^* by contracting a subset of inner edges or equivalently if and only if $\mathcal{H}(T) \subseteq \mathcal{H}(T^*)$.

Lemma 1. *Let T^* be a refinement of T and $u^* v^* \in E(T^*)$. Then there is a unique vertex $w \in V(T)$ such that $L(T(w)) \in \mathcal{H}(T)$ is inclusion-minimal in $\mathcal{H}(T)$ with the property that $L(T^*(v^*)) \subsetneq L(T(w))$. In particular, if $\mathrm{lca}_{T^*}(x, y) = u^*$, then $\mathrm{lca}_T(x, y) = w$.*

Proof. Let $u^* v^* \in E(T^*)$. Since $\mathcal{H}(T) \subseteq \mathcal{H}(T^*)$, $L(T) = L(T^*) \in \mathcal{H}(T)$ and v^* is not the root of T^*, there is a unique inclusion-minimal $A \in \mathcal{H}(T)$ with $L(T^*(v^*)) \subsetneq A$, which corresponds to a unique vertex $w \in V(T)$ that satisfies $L(T(w)) = A$. In the following, we denote with $w^* \in V(T^*)$ the unique vertex that satisfies $A = L(T^*(w^*))$, which exists since $A \in \mathcal{H}(T) \subseteq \mathcal{H}(T^*)$. Now let $x, y \in L(T)$ be two leaves with $\mathrm{lca}_{T^*}(x, y) = u^*$. From $v^* \prec_{T^*} u^*$, we obtain $L(T^*(v^*)) \subsetneq L(T^*(u^*))$ and $L(T^*(u^*)) \subseteq L(T^*(w^*)) = L(T(w))$. Hence, we have $L(T^*(u^*)) \subseteq L(T(w))$, which implies $x, y \in L(T(w))$ and thus also $z := \mathrm{lca}_T(x, y) \preceq_T w$. Denote by $z^* \in V(T^*)$ the unique vertex in T^* with $L(T^*(z^*)) = L(T(z))$. Since $z \preceq_T w$, it satisfies $L(T^*(z^*)) \subseteq L(T^*(w^*))$. Since $x, y \in L(T^*(z^*)) \cap L(T^*(u^*)) \neq \emptyset$, we either have $L(T^*(u^*)) \subseteq L(T^*(z^*))$ or $L(T^*(z^*)) \subsetneq L(T^*(u^*))$. In the second case, we obtain $\mathrm{lca}_{T^*}(x, y) \preceq_{T^*} z^* \prec_{T^*} u^*$, a contradiction to $\mathrm{lca}_{T^*}(x, y) = u^*$. In the first case, we have $L(T^*(v^*)) \subsetneq L(T^*(u^*)) \subseteq L(T(z)) \subseteq L(T(w))$. Due to inclusion minimality of $L(T(w))$ we have $L(T(z)) = L(T(w))$. Thus $\mathrm{lca}_T(x, y) = z = w$. ☐

Lemma 1 ensures that, for every $u^* \in V^0(T^*)$, there is a unique $w \in V(T)$ such that $\mathrm{lca}_T(x, y) = w$ for all $(x, y) \in L_{u^*}^{T^*}$, and thus $L_{u^*}^{T^*} \subseteq L_w^T$. Thus we have

Corollary 1. *If T^* is a refinement of T, then the partition $\mathcal{L}(T^*)$ is a refinement of $\mathcal{L}(T)$.*

Symbolic Ultrametrics. We write $L^{(2)} := \{(x, y) \mid x, y \in L, x \neq y\}$ for the "off-diagonal" pairs of leaves and let M be a finite set.

Definition 1. *A tree T with leaf set L and labeling $t : V^0(T) \to M$ of its inner vertices* explains *a map $\delta : L^{(2)} \to M$ if $t(\mathrm{lca}(x, y)) = \delta(x, y)$ for all distinct $x, y \in L$.*

Such a map must be symmetric since $\mathrm{lca}_T(x,y) = \mathrm{lca}_T(y,x)$ for all $x,y \in L$. A shown in [2], a map $\delta : L^{(2)} \to M$ can be explained by a labeled tree (T,t) if and only if δ is a *symbolic ultrametric*, i.e., iff, for all pairwise distinct $u,v,x,y \in L$ holds (i) $\delta(x,y) = \delta(y,x)$ (symmetry), (ii) $\delta(x,y) = \delta(y,u) = \delta(u,v) \neq \delta(y,v) = \delta(x,v) = \delta(x,u)$ is never satisfied (co-graph property), and (iii) $|\{\delta(u,v), \delta(u,x), \delta(v,x)\}| \leq 2$ (exclusion of rainbow triangles). In this case, there exists a unique least-resolved tree (T_δ, t_δ) (that explains δ) with a discriminating vertex labeling t_δ, i.e., $t_\delta(x) \neq t_\delta(y)$ for all $xy \in E(T_\delta)$ [2,9]. This tree (T_δ, t_δ) is also called a discriminating representation of δ [2].

The construction of symbolic ultrametrics could also be extended to maps $\tilde\delta : L^{(2)} \to 2^M$, i.e., to allow multiple labels at each vertex. However, this does not introduce anything new. To see this, we note that the sets of vertex pairs L_v^T that share the same last common ancestor are pairwise disjoint. In particular, $\tilde\delta$ thus must be a fixed element in 2^M on each L_v^T, $v \in V^0$, and thus we think of the images $\tilde\delta(x,y)$ simply as single labels "associated to" elements in 2^M rather than sets of labels.

Lemma 2. *Let $\delta : L^{(2)} \to M$ be a symbolic ultrametric with least-resolved tree (T_δ, t_δ). Then there is a map $t : V(T) \to M$ such that (T,t) explains δ if and only if T is a refinement of T_δ. In this case, the map t is uniquely determined by T and δ.*

Proof. Suppose (T,t) explains δ and let $e = vu \in E(T)$ be an edge with $t(u) = t(v)$ and $u \prec v$. Note that both u and v must be inner vertices. Let T/e denote the tree obtained from T by contracting the edge e, i.e., removing e from T and identifying u and v. We will keep the vertex v in T/e as placeholder for the identified vertices u and v. By construction, T/e has the clusters $\mathcal{H}(T/e) = \mathcal{H}(T) \setminus \{L(T(u))\}$. Set $t_{T/e}(x) = t(x)$ for all $x \in V^0(T) \setminus \{u\}$. Clearly, v is the unique vertex in T/e such that $L((T/e)(v))$ is inclusion-minimal with property $L(T(u')) \subsetneq L((T/e)(v))$ for any $u' \in \mathrm{child}_T(u)$. Therefore, by Lemma 1, $\mathrm{lca}_T(x,y) = u$ implies $\mathrm{lca}_{T/e}(x,y) = v$, and thus, we have $t(\mathrm{lca}_T(x,y)) = t_{T/e}(\mathrm{lca}_{T/e}(x,y))$ for all $(x,y) \in L^{(2)}$, and thus $(T/e, t_{T/e})$ explains δ. Stepwise contraction of all edges whose endpoints have the same label eventually results in a tree T' and a map t' such that $t'(x) \neq t'(y)$ for all edges of T'. Thus (T',t') coincides with the unique discriminating representation of δ, i.e., $(T',t') = (T_\delta, t_\delta)$. By construction, T is a refinement of T_δ.

Conversely, let δ be a symbolic ultrametric with (unique) discriminating representation (T_δ, t_δ) and let T be a refinement of T_δ. By Corollary 1, $\mathcal{L}(T)$ is a refinement $\mathcal{L}(T_\delta)$. Hence, the map $t : V^0(T) \to M$ specified by $t(\mathrm{lca}_T(x,y)) := t_\delta(\mathrm{lca}_{T_\delta}(x,y))$ for all $(x,y) \in L^{(2)}$ is well-defined. By construction, therefore, (T,t) explains δ. In particular, therefore, every refinement T of T_δ admits a vertex labeling t such that (T,t) explains δ. The choice of t is unique since every inner vertex of a phylogenetic tree is the last common ancestor of at least one pair of vertices, and thus no relabeling of an inner vertex preserves the property that the resulting tree explains δ. $\square$

Fitch Maps encode directional events along edges of T, such as horizontal gene transfer.

Definition 2. *A tree T with edge labeling $\lambda : E(T) \to 2^N$, with finite N, explains a map $\varepsilon : L^{(2)} \to 2^N$ if for all $k \in N$ holds: $k \in \varepsilon(x,y)$ iff $k \in \lambda(e)$ for some edge along the unique path in T that connects $\mathrm{lca}_T(x,y)$ and y.*

A map $\varepsilon : L^{(2)} \to 2^N$ that is explained by a tree (T, λ) in this manner is a *Fitch map* [12]. A Fitch map is called *monochromatic* if $|N| = 1$. Like symbolic ultrametrics, Fitch maps are explained by unique least-resolved trees. The key construction is provided by the sets $U_{\neg m}[y] := \{x \in L \setminus \{y\} \mid m \notin \varepsilon(x,y)\} \cup \{y\}$ for $y \in L$ and $m \in N$. Let us write $\mathcal{N}_\varepsilon := \{U_{\neg m}[y] \mid y \in L, m \in N\}$. Then ε is a Fitch map if and only if (i) $\mathcal{N}_\varepsilon$ is hierarchy-like, i.e., $A \cap B \in \{A, B, \emptyset\}$ for all $A, B \in \mathcal{N}_\varepsilon$ and (ii) $|U_{\neg m}[y']| \leq |U_{\neg m}[y]|$ for all $y \in L$, $m \in N$, and $y' \in U_{\neg m}[y]$ [12, Thm. 3.11].

Fitch maps allow some freedom in distributing labels on the edge set. The precise notion of "least-resolved" thus refers to the fact that it is neither possible to contract edges nor to remove subsets of labels from an edge. The unique least-resolved tree for a Fitch map ε, called the ε-tree $(T_\varepsilon, \lambda_\varepsilon)$, is determined by the hierarchy $\mathcal{H}(T_\varepsilon) = \mathcal{N}_\varepsilon \cup \{L\} \cup \{\{x\} \mid x \in L\}$ and the labeling $\lambda_\varepsilon(\mathrm{parent}_{T_\varepsilon}(v), v) := \{m \in N \mid \exists y \in L \text{ s.t. } L(T_\varepsilon(v)) = U_{\neg m}[y]\}$ for all $e = \{\mathrm{parent}_{T_\varepsilon}(v), v\} \in E(T_\varepsilon)$ [12, Thm. 4.4].

Let (T, λ) and (T', λ') be two edge-labeled trees on the same leaf set and with $\lambda : E(T) \to 2^N$ and $\lambda' : E(T') \to 2^N$. Then (T, λ) is a refinement of (T', λ'), in symbols $(T', \lambda') \leq (T, \lambda)$ if (i) $\mathcal{H}(T') \subseteq \mathcal{H}(T)$ and (ii) if $L(T(v)) = L(T'(v'))$, then $\lambda'(\mathrm{parent}_{T'}(v'), v') \subseteq \lambda(\mathrm{parent}_T(v), v)$.

Proposition 1. *[12, Prop. 4.3, Thm. 4.4] If (T, λ) explains ε, then $(T_\varepsilon, \lambda_\varepsilon) \leq (T, \lambda)$. Furthermore, $(T_\varepsilon, \lambda_\varepsilon)$ is the unique least-resolved tree that explains ε. In particular, $(T_\varepsilon, \lambda_\varepsilon)$ minimizes $\ell_{\min} := \sum_{e \in E(T_\varepsilon)} |\lambda_\varepsilon(e)|$.*

Lemma 3. *Let $\varepsilon : L^{(2)} \to 2^N$ be a Fitch map with least-resolved tree $(T_\varepsilon, \lambda_\varepsilon)$. Then there exists an edge labeling $\lambda : E(T) \to 2^N$ such that (T, λ) explains ε if and only if T is a refinement of T_ε.*

Proof. Suppose (T, λ) explains ε. By Proposition 1, this implies $(T_\varepsilon, \lambda_\varepsilon) \leq (T, \lambda)$, i.e., T is a refinement of T_ε. Conversely, let ε be a Fitch map with least-resolved tree $(T_\varepsilon, \lambda_\varepsilon)$ and let T be a refinement of T_ε. Define, for all edges $\{\mathrm{parent}_T(v), v\} \in E(T)$, the edge labeling

$$\lambda(\{\mathrm{parent}_T(v), v\}) := \begin{cases} \lambda_\varepsilon(\mathrm{parent}_{T_\varepsilon}(v'), v') & \text{if } L(T(v)) = L(T_\varepsilon(v')), \\ \emptyset & \text{otherwise.} \end{cases} \tag{1}$$

The map λ is well-defined, since there is at most one $v' \in V(T_\varepsilon)$ with $L(T(v)) = L(T_\varepsilon(v'))$.

Claim. (T, λ) and $(T_\varepsilon, \lambda_\varepsilon)$ explain the same Fitch map ε.

By assumption, $(T_\varepsilon, \lambda_\varepsilon)$ explains ε. Let $(a, b) \in L^{(2)}$, $k \in N$, and let ε' be the

Fitch map explained by (T, λ). First, suppose $k \in \varepsilon(a, b)$, i.e., there is an edge $e' = \{\text{parent}_{T_\varepsilon}(w'), w'\}$ with $k \in \lambda_\varepsilon(e')$ such that $w' \prec_{T_\varepsilon} \text{lca}_{T_\varepsilon}(a, b)$ by the definition of Fitch maps. We have $a \notin L(T_\varepsilon(w'))$. Since T is a refinement of T_ε, there is a vertex $w \in V(T)$ with $L(T(w)) = L(T_\varepsilon(w'))$. In particular, therefore, $\lambda(\{\text{parent}_T(w), w\}) = \lambda_\varepsilon(e')$. This together with the fact that $a \notin L(T_\varepsilon(w')) = L(T(w))$ immediately implies $k \in \varepsilon'(a, b)$. Now suppose $k \in \varepsilon'(a, b)$. Hence, there is an edge $e = \{\text{parent}_T(v), v\}$ with $v \prec_T \text{lca}_T(a, b)$ and $k \in \lambda(e)$. By construction of λ, the latter implies that there is a vertex $v' \in V(T_\varepsilon)$ with $L(T(v)) = L(T_\varepsilon(v'))$ and, in particular, $k \in \lambda_\varepsilon(\text{parent}_{T_\varepsilon}(v'), v')$. The latter together with $a \notin L(T(v)) = L(T_\varepsilon(v'))$ implies that $k \in \varepsilon(a, b)$. Since $(a, b) \in L^{(2)}$ and $k \in N$ were chosen arbitrarily, we conclude that $\varepsilon = \varepsilon'$, and thus, (T, λ) also explains ε.   □

The labeling λ defined in Eq. (1) satisfies $\ell_{\min} = \sum_{e \in T(e)} |\lambda(e)|$ by construction and Proposition 1. Furthermore, we observe that (T^*, λ^*) is obtained from (T, λ) by contracting only edges with $\lambda(e) = \emptyset$. More precisely, e is contracted if and only if e is an inner edge with $\lambda(e) = \emptyset$. This implies

Corollary 2. *Suppose (T, λ') explains the Fitch map ε. Then $\lambda : E(T) \to 2^N$ given by Eq. (1) is the unique labeling such that (T, λ) explains ε and $\sum_{e \in E(T)} |\lambda(e)| = \ell_{\min}$.*

Proof. Suppose (T, λ'') explains ε and $\sum_{e \in E(T)} |\lambda''(e)| = \ell_{\min}$. By Proposition 1, we have $(T_\varepsilon, \lambda_\varepsilon) \leq (T, \lambda'')$ and thus $\lambda_\varepsilon(\text{parent}_{T_\varepsilon}(v'), v') \subseteq \lambda''(\text{parent}_T(v), v)$ if $L(T_\varepsilon(v')) = L(T(v))$. Since, moreover, $\lambda_\varepsilon(\text{parent}_{T_\varepsilon}(v'), v') = \lambda(\text{parent}_T(v), v)$ if $L(T_\varepsilon(v')) = L(T(v))$ by Eq. (1), minimality of λ'' implies $\lambda'' = \lambda$.   □

3   Tree-Like Pairs of Maps

Symbolic ultrametrics and Fitch maps on $L^{(2)}$ derive from trees in very different ways by implicitly leveraging information about inner vertices and edges of the *a priori* unknown tree. It is of interest, therefore, to know when they are consistent in the sense that they can be simultaneously explained by a tree.

Definition 3. *An ordered pair (δ, ε) of maps $\delta : L^{(2)} \to M$ and $\varepsilon : L^{(2)} \to 2^N$ is tree-like if there is a tree T endowed with a vertex labeling $t : V^0(T) \to M$ and edge labeling $\lambda : L^{(2)} \to 2^N$ such that (T, t) explains δ and (T, λ) explains ε.*

Naturally, we ask when (δ, ε) is explained by a vertex and edge labeled tree (T, t, λ), i.e., when (δ, ε) is a tree-like pair of maps on $L^{(2)}$. Furthermore, we ask whether a tree-like pair of maps is again explained by a unique least-resolved tree (T^*, t^*, λ^*).

Theorem 1. *Let $\delta : L^{(2)} \to M$ and $\varepsilon : L^{(2)} \to 2^N$. Then (δ, ε) is tree-like if and only if*

1. δ is a symbolic ultrametric.
2. ε is a Fitch map.
3. $\mathcal{H}^* := \mathcal{H}(T_\delta) \cup \mathcal{H}(T_\varepsilon)$ is a hierarchy.

In this case, there is a unique least-resolved vertex and edge labeled tree (T^*, t^*, λ^*) explaining (δ, ε). The tree T^* is determined by $\mathcal{H}(T^*) = \mathcal{H}^*$, the vertex labeling t^* is uniquely determined by t_δ and the edge labeling λ^* with minimum value of $\sum_{e \in E(T^*)} |\lambda^*(e)|$ is uniquely determined by λ_ε.

Proof. Suppose (δ, ε) is tree-like, i.e., there is a tree (T, t, λ) such that (T, t) explains δ and (T, λ) explains ε. Thus δ is a symbolic ultrametric and ε is a Fitch map. Furthermore, T is a refinement of least-resolved trees T_δ and T_ε because of the uniqueness of these least-resolved trees, and we have $\mathcal{H}(T_\delta) \subseteq \mathcal{H}(T)$ and $\mathcal{H}(T_\varepsilon) \subseteq \mathcal{H}(T)$ and thus $\mathcal{H}^* \subseteq \mathcal{H}(T)$. Since $\mathcal{H}(T)$ is a hierarchy and the subset $\mathcal{H}^*$ contains both L and all singletons $\{x\}$ with $x \in L$, $\mathcal{H}^*$ is a hierarchy.

Conversely, suppose conditions (1), (2), and (3) are satisfied. The first two conditions guarantee the existence of the least-resolved tree (T_δ, t_δ) and $(T_\varepsilon, \lambda_\varepsilon)$ explaining δ and ε, respectively. Thus $\mathcal{H}^* = \mathcal{H}(T_\delta) \cup \mathcal{H}(T_\varepsilon)$ is well-defined. Condition (3) stipulates that $\mathcal{H}^*$ is a hierarchy and thus there is a unique tree T^* such that $\mathcal{H}(T^*) = \mathcal{H}^*$, which by construction is a refinement of both T_δ and T_ε. By Lemmas 2 and 3, T^* can be equipped with a vertex-labeling t^* and an edge-labeling λ^* such that (T^*, t^*) explains δ and (T^*, λ^*) explains ε, respectively. Thus (δ, ε) is tree-like.

We now show that (T^*, t^*, λ^*) is least-resolved w.r.t. (δ, ε) and thus that for every $e \in E(T^*)$, the tree $T' := T^*/e$ does not admit a vertex labeling $t' : V^0(T') \to M$ and an edge-labeling $\lambda' : E(T') \to 2^N$ such that (T', t', λ') explains (δ, ε). Let $e = \{\text{parent}_{T^*}(v), v\} \in E(T^*)$. Hence, $L(T^*(v)) \in \mathcal{H}(T^*)$. If $v \in L(T^*)$, then we have $L(T^*) \neq L(T')$ and the claim trivially holds. Thus suppose that $v \in V^0(T)$ in the following. Since the edge e is contracted in T', we have $\mathcal{H}(T') = \mathcal{H}(T^*) \setminus \{L(T(v))\}$ and thus, $\mathcal{H}(T_\delta) \not\subseteq \mathcal{H}(T')$ or $\mathcal{H}(T_\varepsilon) \not\subseteq \mathcal{H}(T')$. Thus T' is not a refinement of T_δ or T_ε. By Lemmas 2 and 3, this implies that there is no t' such that (T', t') explains δ or no λ' such that (T', λ') explains ε, respectively. Thus (T^*, t^*, λ^*) is least-resolved w.r.t. (δ, ε).

It remains to show that (T^*, t^*, λ^*) is unique. Since T^* is uniquely determined by $\mathcal{H}^*$, it suffices to show that the labeling of T^* is unique. This, however, follows immediately from Lemma 2 and Corollary 2, respectively. $\square$

We note that every refinement T of the least-resolved tree (T^*, t^*, λ^*) admits a vertex labeling $t : V^0(T) \to M$ and an edge labeling $\lambda : E(T) \to 2^N$ such that (T, t, λ) explains (δ, ε).

Theorem 2. *Given two maps $\delta : L^{(2)} \to M$ and $\varepsilon : L^{(2)} \to 2^N$ it can be decided in $O(|L|^2|N|)$ whether (δ, ε) is tree-like. In the positive case, the unique least-resolved tree (T^*, t^*, λ^*) can be obtained with the same effort.*

Proof. Based on Theorem 1, a possible algorithm consists of three steps: (i) check whether δ is a symbolic ultrametric, (ii) check whether ε is a Fitch map and, if both statements are true, (iii) compute $\mathcal{H}^* := \mathcal{H}(T_\delta) \cup \mathcal{H}(T_\varepsilon)$ and use this information to compute the unique least-resolved vertex and edge labeled tree (T^*, t^*, λ^*). By [12, Thm. 6.2], the decision whether ε is a Fitch map and, in the positive case, the construction of the least-resolved tree $(T_\varepsilon, \lambda_\varepsilon)$ can be achieved in $O(|L|^2 |N|)$ time. Moreover, it can be verified in $O(|L|^2)$ whether or not a given map δ is a symbolic ultrametric, and, in the positive case, the discriminating tree (T_δ, t_δ) can be computed within the same time complexity (cf. [13, Thm. 7]). The common refinement T with $\mathcal{H}(T) = \mathcal{H}(T_\delta) \cup \mathcal{H}(T_\varepsilon)$ can be computed in $O(|L|)$ time using LinCR [18].

The edge labels λ^* are then carried over from $(T_\varepsilon, \lambda_\varepsilon)$ using the correspondence between $u^* v^* \in E(T^*)$ and $uv \in E(T_\varepsilon)$ iff $L(T^*(v^*)) = L(T_\varepsilon(v))$, otherwise $\lambda^*(\{u^*, v^*\}) = \emptyset$. This requires $O(|L| \cdot |N|)$ operations. The vertex labels can then be assigned by computing, for all $(x, y) \in L^{(2)}$, the vertex $v = \mathrm{lca}_{T^*}(x, y)$ and assigning $t^*(v) = \delta(x, y)$ in quadratic time using a fast last common ancestor algorithm [7]. Thus we arrive at a total performance bound of $O(|L|^2 |N|)$.   □

4 Tree-Like Pairs of Maps with Constraints

One interpretation of tree-like pairs of maps (δ, ε) is to consider δ as the orthology relation and ε as the xenology relation. In such a setting, certain vertex labels $t(v)$ preclude some edge labels $\lambda(\{v, u\})$ with $u \prec v$. For example, a speciation vertex cannot be the source of a horizontal transfer edge. We use the conventional notations $t(u) = \bullet$ and $t(v) = \square$ for speciation and duplication vertices [6], respectively, set $t(u) = \triangle$ for a third vertex type, and consider the monochromatic Fitch map $\varepsilon \colon L^{(2)} \to \{\emptyset, \mathbb{I}\}$. Thus, we require that $\lambda(\{v, u\}) = \mathbb{I}$ and $u \prec_T v$ implies $t(v) = \triangle$ [1,17,20]. This condition simply states that neither a speciation nor a gene duplication is the source of a horizontal transfer.

In [17], we considered evolutionary scenarios that satisfy another rather stringent observability condition:

(C) For every $v \in V^0(T)$, there is a child $u \in \mathrm{child}_T(v)$ such that $\lambda(\{v, u\}) = \emptyset$.

We call a Fitch map λ that satisfies (C) a *type-C Fitch map*. In this case, for every $v \in V^0(T)$, there is a leaf $x \in L(T(v))$ such that $\lambda(e) = \emptyset$ for all edges along the path from v to x. As an immediate consequence of (C), we observe that, given $|L| \geq 2$, for every $x \in L$ there is a $y \neq x$ such that $\varepsilon(x, y) = \emptyset$. This condition is not sufficient, however, as the following example shows. Consider the tree $((x, y), (x', y'))$ in Newick notation, with the edges in the two cherries (x, y) and (x', y') being labeled with $\emptyset$, and two $\mathbb{I}$-labeled edges incident to the root. Then, for every $z \in L$, we have $\varepsilon(z, z') = \emptyset$, where z' is the sibling of z, but condition (C) is not satisfied. In a somewhat more general setting, we formalize these two types of labeling constraints as follows:

Definition 4. *Let $\delta \colon L^{(2)} \to M$ and $\varepsilon \colon L^{(2)} \to 2^N$ be two maps and $M_\emptyset \subseteq M$. Then, (δ, ε) is $M_\emptyset$-tree-like if there is a tree (T, t, λ) that explains (δ, ε) and the labeling maps $t \colon V^0(T) \to M$ and $\lambda \colon E(T) \to 2^N$ satisfy (C) and*

(C1) *If $t(v) \in M_\emptyset$, then $\lambda(\{v, u\}) = \emptyset$ for all $u \in \mathrm{child}_T(v)$.*

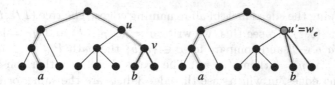

Fig. 1. Effect of an edge contraction on paths in T. All paths traversing the contracted edge $e = uv$ in T correspond to paths in T/e in which e is contracted. All other path remain unchanged. Furthermore $w_e = u'$, i.e., the edge contraction corresponds to the deletion of $L(T(v))$ from $\mathcal{H}(T)$.

Hence, $M_\emptyset$ puts extra constraints to the vertex and edge labels on trees that satisfy (C) and explain (δ, ε). Note, an $\emptyset$-tree-like ($M_\emptyset = \emptyset$) must only satisfy (C) and (C1) can be omitted.

Theorem 3. *Let $\delta : L^{(2)} \to M$ and $\varepsilon : L^{(2)} \to 2^N$ be two maps and $M_\emptyset \subseteq M$. Then, (δ, ε) is $M_\emptyset$-tree-like if and only if (δ, ε) is tree-like and its least-resolved tree (T^*, t^*, λ^*) satisfies (C) and (C1).*

Proof. If (δ, ε) is tree-like and its least-resolved tree (T^*, t^*, λ^*) satisfies (C) and (C1), then (δ, ε) is $M_\emptyset$-tree-like by definition. For the converse, suppose (δ, ε) is $M_\emptyset$-tree-like and let (T, t, λ) be a vertex and edge labeled tree that explains (δ, ε) and satisfies (C) and (C1).

Let λ' be the edge labeling for T as specified in Eq. (1) where $(T_\varepsilon, \lambda_\varepsilon)$ is replaced by (T^*, λ^*). By the arguments in the proof of Lemma 3, (T, λ') still explains ε and hence, (T, t, λ') explains (δ, ε). Moreover, since $\ell_{\min} := \sum_{e \in E(T^*)} |\lambda^*(e)|$ and by construction of λ', we have $\ell_{\min} = \sum_{e \in E(T)} |\lambda'(e)|$. Since $(T^*, \lambda^*) \leq (T, \lambda')$, it must hold $\lambda_\varepsilon(e') \subseteq \lambda'(e)$ for all $e' = \mathrm{parent}_{T_\varepsilon}(v')v' \in E(T_\varepsilon)$ and $e = \mathrm{parent}_T(v)v \in E(T)$ with $L(T(v)) = L(T_\varepsilon(v'))$. Since λ' is minimal by construction, we have $\lambda_\varepsilon(e') = \lambda'(e)$ for all corresponding edges e and e'. In particular, it must hold that $\lambda(e) = \emptyset$ implies $\lambda'(e) = \emptyset$ for all $e \in E(T)$. To see this, assume for contradiction there is some edge $e = uv \in E(T)$ with $\lambda(e) = \emptyset$ but $\lambda'(e) \neq \emptyset$. Since (T, λ) satisfies (C), there is a path from u to some leaf $y \in L(T)$ that consists of edges f with label $\lambda(f) = \emptyset$ only and that contains the edge e. Hence, for $x \in L(T(u)) \backslash L(T(v))$, we have $\mathrm{lca}_T(x, y) = u$ and thus, $\varepsilon(x, y) = \emptyset$. However, since we assume that $\lambda'(e) = N' \neq \emptyset$, we obtain $N' \subseteq \varepsilon(x, y) \neq \emptyset$; a contradiction. Now it is easy to verify that (T, t, λ') still satisfies (C) and (C1).

Now consider edge contractions, illustrated in Fig. 1. To obtain T^* we are only allowed to contract edges $e = uv \in E(T)$ that satisfy $t(u) = t(v)$ and $\lambda'(e) = \emptyset$. The latter follows from the fact that edges uv with $t(u) \neq t(v)$ cannot be contracted without losing the information of at least one of the labels $t(u)$ or $t(v)$ and minimality of λ', since otherwise the labels $\lambda'(e)$ do not contribute to the explanation of the Fitch map and thus would have been removed in the construction of λ'. For such an edge e, the tree $(T/e, t_{T/e}, \lambda'_{T/e})$ is obtained by contracting the edge $e = uv$ to a new vertex w_e and assigning $t_{T/e}(w_e) = t(v) =$

$t(u)$ and keeping the edge labels of all remaining edges. The tree $(T/e, t_{T/e}, \lambda'_{T/e})$ then explains (δ, ε). To see this, we write $y := \mathrm{lca}_T(a, b)$ and $y' := \mathrm{lca}_{T/e}(a, b)$ for distinct $a, b \in L$ and compare for $c \in \{a, b\}$ the path P_{yc} in T and $P'_{y'c}$ in T/e. If $y = u$ or $y = v$ then $y' = w_e$. The paths therefore either consist only of corresponding edges, in which case the edge labels are the same, or they differ exactly by the contraction of e. The latter does not affect the explanation of $\varepsilon(a, b)$ because $\lambda'(e) = \emptyset$. Since $t(u) = t(v)$, contraction of uv also does not affect δ.

In particular, therefore, neither u nor v is a leaf, i.e., e is an inner edge. Condition (C) is trivially preserved under contraction of inner edges. Suppose $t(v) = t(u) \in M_\emptyset$ and thus $t_{T/e}(w_e) \in M_\emptyset$. Since (T, t, λ') satisfies (C1) we have $\lambda'(\{v, u'\}) = \lambda'(\{u, u''\}) = \emptyset$ for all $u' \in \mathrm{child}_T(v)$ and all $u'' \in \mathrm{child}_T(u)$ and thus after contracting e it holds that $\lambda'_{T/e}(w_e, w') = \emptyset$ for all $w' \in \mathrm{child}_{T/e}(w_e) = \mathrm{child}_T(v) \uplus \mathrm{child}_T(u)$. Otherwise, $t(u) = t(v) \notin M_\emptyset$ and thus by construction $t_{T/e}(w_e) \notin M_\emptyset$. In summary, $(T/e, t', \lambda)$ satisfies (C) and (C1). Repeating this coarse graining until no further contractible inner edges are available results in the unique least-resolved tree (T^*, t^*, λ^*).   □

Since the unique least-resolved tree (T^*, t^*, λ^*) can be computed in quadratic time by Theorem 2, and it suffices by Theorem 3 to check (C) and (C1) for (T^*, t^*, λ^*), the same performance bound applies to the recognition of constrained tree-like pairs of maps.

We note that an analogous result holds if only (C) or only (C1) is required for (T, t, λ). Furthermore, one can extend (C1) in such a way that for a set $\mathcal{Q}$ of pairs (q, m), with $q \in M$ and $m \in N$, of labels that are incompatible at a vertex v and an edge vv' with $v' \in \mathrm{child}_T(v)$. The proof of Theorem 3 still remains valid since also in this case no forbidden combinations of vertex an edge colors can arise from contracting an edge $e = uv$ with $t(u) = t(v)$. In the special case $\delta(x, y) = 1 \notin M_\emptyset$ for all $(x, y) \in L^{(2)}$, one obtains $t^*(u) = 1$ for all $u \in V(T^*)$ and thus $(T^*, \lambda^*) = (T_\varepsilon, \lambda_\varepsilon)$ and (C1) imposes no constraint. Hence, Theorem 3 specializes to

Corollary 3. *A Fitch map ε is type-C if and only if its least-resolved tree $(T_\varepsilon, \lambda_\varepsilon)$ satisfies (C).*

In [17] a stronger version of condition (C) has been considered:

(C2) If $\lambda(\{v, u\}) \neq \emptyset$ for some $u \in \mathrm{child}_T(v)$, then $\lambda(\{v, u'\}) = \emptyset$ for *all* $u' \in \mathrm{child}_T(v) \setminus \{u\}$.

This variant imposes an additional condition on the edges $e = uv$ that can be contracted. More precisely, an inner edge of (T, t, λ) can be contracted without losing the explanation of (δ, ε) and properties (C1) and (C2) if and only if (i) $t(u) = t(v)$, (ii) $\lambda(e) = \emptyset$ and (iii) at most one of the edges uu', $u' \in \mathrm{child}_T(u)$ and vv', $v' \in \mathrm{child}_T(v)$ has a non-empty label. Now consider two consecutive edges uv and vw with $t(u) = t(v) = t(w)$, $\lambda(\{u, v\}) = \lambda(\{v, w\}) = \emptyset$ and suppose there is $u' \in \mathrm{child}_T(u)$ with $\lambda(\{u, u'\}) \neq \emptyset$, $w' \in \mathrm{child}_T(w)$ with $\lambda(\{w, w'\}) \neq \emptyset$, and

$\lambda(\{v, v'\}) = \emptyset$ for all $v' \in \text{child}_T(v)$. Then one can contract either uv or vw but not both edges. Thus least-resolved trees explaining (δ, ε) and satisfying (C1) and (C2) are no longer unique.

5 Concluding Remark

Here we have shown that symbolic ultrametrics and Fitch maps can be combined by the simple and easily verified condition that $\mathcal{H}(T_\delta) \cup \mathcal{H}(T_\varepsilon)$ is again a hierarchy (Theorem 1), i.e., that the two least-resolved trees have a common refinement. The least-resolved tree (T^*, t^*, λ^*) that simultaneously explains both δ and ε is unique in this case and can be computed in quadratic time if the label set N is bounded and $O(|L|^2|N|)$ time in general. The closely related problem of combining a hierarchy and *symmetrized Fitch maps*, defined by $m \in \varepsilon(x, y)$ iff there is an edge e with $m \in \lambda(e)$ along the path from x to y [10], is NP-complete [14]. It appears that the main difference is the fact that symmetrized Fitch maps do not have a unique least-resolved tree as explanation. The distinction between much simpler problems in the directed setting and hard problems in the undirected case is also reminiscent of the reconciliation problem for trees, which are easy for rooted trees and hard for unrooted trees, see e.g. [3].

We have also seen that certain restrictions on the Fitch maps that are related to the "observability" of horizontal transfer do not alter the complexity of the problem. These observability conditions are defined in terms of properties of the explaining trees, raising the question whether these constraints also have a natural characterization as properties of the Fitch maps. On a more general level, both symbolic ultrametrics and Fitch maps arise from evolutionary scenarios comprising an embedding of the gene tree T into a species tree, with labeling functions t and λ on T encoding event-types and distinctions in the evolutionary fate of offsprings, respectively. Here we have focused entirely on gene trees with given labels. The embeddings into species trees are known to impose additional constraints [8,16].

Acknowledgments. This work was supported in part by the *Deutsche Forschungsgemeinschaft* (STA 850/51-2).

References

1. Bansal, M.S., Alm, E.J., Kellis, M.: Efficient algorithms for the reconciliation problem with gene duplication, horizontal transfer and loss. Bioinformatics **28**, i283–i291 (2012). https://doi.org/10.1093/bioinformatics/bts225
2. Böcker, S., Dress, A.W.M.: Recovering symbolically dated, rooted trees from symbolic ultrametrics. Adv. Math. **138**, 105–125 (1998). https://doi.org/10.1006/aima.1998.1743
3. Bryant, D., Lagergren, J.: Compatibility of unrooted phylogenetic trees is FPT. Theor. Comput. Sci. **351**, 296–302 (2006). https://doi.org/10.1016/j.tcs.2005.10.033

4. Fitch, W.: Homology: a personal view on some of the problems. Trends Genet. **16**, 227–231 (2000). https://doi.org/10.1016/S0168-9525(00)02005-9
5. Geiß, M., Anders, J., Stadler, P.F., Wieseke, N., Hellmuth, M.: Reconstructing gene trees from Fitch's xenology relation. J. Math. Biol. **77**(5), 1459–1491 (2018). https://doi.org/10.1007/s00285-018-1260-8
6. Geiß, M., et al.: Best match graphs and reconciliation of gene trees with species trees. J. Math. Biol. **80**(5), 1459–1495 (2020). https://doi.org/10.1007/s00285-020-01469-y
7. Harel, D., Tarjan, R.: Fast algorithms for finding nearest common ancestors. SIAM J. Comput. **13**, 338–355 (1984). https://doi.org/10.1137/0213024
8. Hellmuth, M.: Biologically feasible gene trees, reconciliation maps and informative triples. Algorithms Mol. Biol. **12**, 23 (2017). https://doi.org/10.1186/s13015-017-0114-z
9. Hellmuth, M., Hernandez-Rosales, M., Huber, K.T., Moulton, V., Stadler, P.F., Wieseke, N.: Orthology relations, symbolic ultrametrics, and cographs. J. Math. Biol. **66**, 399–420 (2013). https://doi.org/10.1007/s00285-012-0525-x
10. Hellmuth, M., Long, Y., Geiß, M., Stadler, P.F.: A short note on undirected Fitch graphs. Art Discrete Appl. Math. **1,** P1.08 (2018). https://doi.org/10.26493/2590-9770.1245.98c
11. Hellmuth, M., Schaller, D., Stadler, P.F.: Compatibility of partitions, hierarchies, and split systems (2021, submitted). arXiv arXiv:2104.14146
12. Hellmuth, M., Seemann, C.R., Stadler, P.F.: Generalized Fitch graphs II: sets of binary relations that are explained by edge-labeled trees. Discrete Appl. Math. **283**, 495–511 (2020). https://doi.org/10.1016/j.dam.2020.01.036
13. Hellmuth, M., Stadler, P.F., Wieseke, N.: The mathematics of xenology: dicographs, symbolic ultrametrics, 2-structures and tree-representable systems of binary relations. J. Math. Biol. **75**(1), 199–237 (2016). https://doi.org/10.1007/s00285-016-1084-3
14. Hellmuth, M., Seemann, C.R., Stadler, P.F.: Generalized Fitch graphs III: symmetrized Fitch maps and sets of symmetric binary relations that are explained by unrooted edge-labeled trees. Discrete Math. Theor. Comput. Sci **23**(1), 13 (2021)
15. Jones, M., Lafond, M., Scornavacca, C.: Consistency of orthology and paralogy constraints in the presence of gene transfers (2017). arXiv arXiv:1705.01240
16. Lafond, M., Hellmuth, M.: Reconstruction of time-consistent species trees. Algorithms Mol. Biol. **15**, 16 (2020). https://doi.org/10.1186/s13015-020-00175-0
17. Nøjgaard, N., Geiß, M., Merkle, D., Stadler, P.F., Wieseke, N., Hellmuth, M.: Time-consistent reconciliation maps and forbidden time travel. Algorithms Mol. Biol. **13**, 2 (2018). https://doi.org/10.1186/s13015-018-0121-8
18. Schaller, D., Hellmuth, M., Stadler, P.F.: A linear-time algorithm for the common refinement of rooted phylogenetic trees on a common leaf set (2021, submitted). arXiv arXiv:2107.00072
19. Semple, C., Steel, M.: Phylogenetics. Oxford University Press, Oxford (2003)
20. Tofigh, A., Hallett, M., Lagergren, J.: Simultaneous identification of duplications and lateral gene transfers. IEEE/ACM Trans. Comput. Biol. Bioinform **8**(2), 517–535 (2011). https://doi.org/10.1109/TCBB.2010.14

ContFree-NGS: Removing Reads from Contaminating Organisms in Next Generation Sequencing Data

Felipe Vaz Peres[1,2] (iD) and Diego Mauricio Riaño-Pachón[2(✉)] (iD)

[1] Federal University of São Carlos, Araras, , São Paulo, Brazil
[2] Computational, Evolutionary and Systems Biology Laboratory, Center for Nuclear Energy in Agriculture, University of São Paulo, Piracicaba, , São Paulo, Brazil
diego.riano@cena.usp.br

Abstract. We present ContFree-NGS, an open source software that removes reads originating from contaminant organisms in your sequencing dataset. The user has to provide a target taxon, and anything that does not belong to this taxon or its descendants will be labelled as contaminant. In order to achieve this, ContFree-NGS exploits results from a taxonomy assignment engine, like Kraken2 or Kaiju.

Keywords: NGS · Contamination · Bioinformatics

1 Introduction

Second and third generation DNA sequencing technologies are powerful tools that are revolutionizing biology. However, results from these technologies often present contamination, which could impact their interpretation [1, 2]. A contaminating sequence is one that does not faithfully represent the genetic information from the biological source organism because it contains one or more sequence segments of foreign origin, and they could cause several problems in downstream analyses. The primary consequences of contamination are time and effort wasted on meaningless analyses, erroneous conclusions drawn about the biological significance of the sequence, misassembly of sequence contigs and false sequence clustering, delay in the release of the sequence in public databases and pollution of public databases [3].

Recently, some tools have been made available that aim to remove sequences from contaminating organisms in next generation sequencing (NGS) datasets. DecontaMiner is a tool to unravel the presence of contaminating sequences in the set of reads that do not map to a reference genome [4]. Conterminator removes contaminating sequences from contigs exploiting a taxonomic assignment file [5]. QC-Blind is an automatic tool to do unsupervised assembly and contig binning to identify and remove putative contaminants [6]. These tools have in common that they either require a reference genome of the source organism, or need to perform assembly prior contaminant detection. Our goal was to develop a simpler tool to remove contaminated sequences directly from unassembled reads, without mapping, exploiting fast k-mer analysis implemented in

© Springer Nature Switzerland AG 2021
P. F. Stadler et al. (Eds.): BSB 2021, LNBI 13063, pp. 65–68, 2021.
https://doi.org/10.1007/978-3-030-91814-9_6

taxonomic assignment engines commonly used in metagenomics. Thus, we developed ContFree-NGS, an open source and very simple filter that removes sequences from contaminating organisms in NGS datasets based on a taxonomic classification file.

2 Implementation

ContFree-NGS was implemented as a single Python v3 (>3.6) script, using the biopython module and the Python Environment for Tree Exploration (ETE). In order to assess contamination, ContFree-NGS exploits a taxonomic assignment file containing the read identifier and a NCBI taxonomic identifier for every sequence in the dataset. This taxonomic classification file can be generated with a taxonomic assignment engine, such as Kraken2 [7] or Kaiju [8]. ContFree-NGS requires that the user provide a target taxon and only reads assigned to this taxon or to its descendants will be regarded as target sequences and further maintained. Sequences not assigned to the target taxon or its descendants will be discarded and sequences that could not be assigned to any taxa will be kept in a separated unclassified file. ContFree-NGS will process the NGS dataset and the taxonomic assignment file in the following way:

1. It creates an indexed database for the sequencing dataset (FastQ format) using the Bio.SeqIO.index_db function with the index stored in a SQLite database;
2. It creates a list with the NCBI identifier for the target taxon and all the identifiers of its descendants according to the NCBI taxonomy database;
3. It iterates over the taxonomy assignment file. If the read was not assigned to any taxa it is saved in a fastq file for unclassified reads. If the read was assigned, it will check if its taxon is found in the list of the target taxon descendants, created in step (ii), if so, will save the read to a fastq file for filtered reads, otherwise the read will be discarded.

As ContFree-NGS exploits the results from a taxonomic assignment engine, users must use the proper switches to achieve an accurate classification, for instance a proper value of the --confidence switch in Kraken2.

3 Evaluation

We evaluated ContFree-NGS on three sugarcane artificially contaminated datasets, A (50.000 paired end reads), B (250.000 paired end reads) and C (1.250.000 paired end reads). In all datasets 80% of the reads came from sugarcane (SRR1774134), 15% came from *Acinetobacter baumanii* (SRR12763742) and 5% came from *Aspergillus fumigatus* (DRR289670). We used Kraken2 for taxonomic assignment. To perform that, we built a Kraken2 custom database containing the following reference libraries: archaea, bacteria, viral, human, fungi, plant, protozoa and the NCBI non-redundant nucleotide database. Then, Kraken2 was run with the confidence set to 0.05, resulting in the following number of classified sequences: dataset A: 25.547, dataset B: 128.396, dataset C: 664.270.

At the confidence level of 0.05 set in Kraken2, ContFree-NGS was able to remove over 99% of the known contaminants in the set of classified reads. We run ContFree-NGS on a high performance computing (HPC) cluster and recorded RAM usage and processing wall time for the three datasets (see Fig. 1).

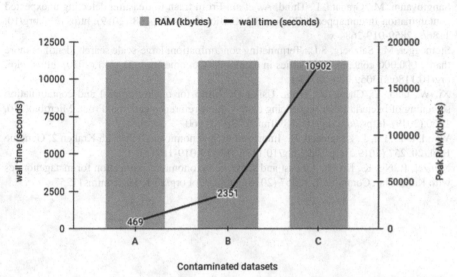

Fig. 1. This figure shows the RAM usage and time consuming to remove contaminants of the three sugarcane artificially contaminated datasets. Memory usage is low and independent of the number of classified sequences and wall time scale rapidly with the number of classified sequences. To reduce time consuming, the end user could split the taxonomy assignment file in several files and process them in parallel. Check our GitHub (https://github.com/labbces/ContFree-NGS) page for more details.

4   Conclusion

ContFree-NGS is a very simple filter and useful tool that removes sequences from contaminating organisms in a NGS dataset.

Funding. This work was supported by "Fundação de Amparo à Pesquisa do Estado de São Paulo (FAPESP)" [grant number 2019/24796-5 to F.V.P] and by "Conselho Nacional de Desenvolvimento Científico e Tecnológico (CNPq)" [grant number 310080/2018-5 to D.M.R-P].

References

1. Park, S.J., Onizuka, S., Seki, M., et al.: A systematic sequencing-based approach for microbial contaminant detection and functional inference. BMC Biol. **17**, 72 (2019). https://doi.org/10.1186/s12915-019-0690-0

2. Goig, G.A., Blanco, S., Garcia-Basteiro, A.L., et al.: Contaminant DNA in bacterial sequencing experiments is a major source of false genetic variability. BMC Biol. **18**, 24 (2020). https://doi.org/10.1186/s12915-020-0748-z
3. National Center for Biotechnology Information 2016: Contamination in Sequence Databases. https://www.ncbi.nlm.nih.gov/tools/vecscreen/contam/. Accessed 6 Oct 2021
4. Sangiovanni, M., Granata, I., Thind, A., et al.: From trash to treasure: detecting unexpected contamination in unmapped NGS data. BMC Bioinform. **20**, 168 (2019). https://doi.org/10.1186/s12859-019-2684-x
5. Steinegger, M., Salzberg, S.L.: Terminating contamination: large-scale search identifies more than 2,000,000 contaminated entries in GenBank. Genome Biol. **21**, 115 (2020). https://doi.org/10.1186/s13059-020-02023-1
6. Xi, W., Gao, Y., Cheng, Z., et al.: Using QC-blind for quality control and contamination screening of bacteria DNA sequencing data without reference genome. Front. Microbiol. **10**, 1560 (2019). https://doi.org/10.3389/fmicb.2019.01560
7. Wood, D.E., Lu, J., Langmead, B.: Improved metagenomic analysis with Kraken 2. Genome Biol. **20**, 257 (2019). https://doi.org/10.1186/s13059-019-1891-0
8. Menzel, P., Ng, K., Krogh, A.: Fast and sensitive taxonomic classification for metagenomics with Kaiju. Nat. Commun. **7**, 11257 (2016). https://doi.org/10.1038/ncomms11257

Deep Learning-Based COVID-19 Diagnostics of Low-Quality CT Images

Daniel Ferber, Felipe Vieira, João Dalben, Mariana Ferraz, Nicholas Sato, Gabriel Oliveira, Rafael Padilha[✉], and Zanoni Dias

Institute of Computing, University of Campinas, Campinas, SP, Brazil
`rafael.padilha@ic.unicamp.br`

Abstract. Mass testing of the population is among the most effective measures to combat the COVID-19 pandemic. Among existing diagnostic methods, deep learning-based solutions have the potential to be affordable, quick and accurate. However, these techniques often rely on high-quality datasets, which are not always available in medical scenarios. In this work, we use convolutional neural networks to diagnose COVID-19 on computed tomography images from the COVIDx-CT dataset [6]. The available scans often present noisy artifacts, originated from sensor- and capturing-related errors, that can negatively impact the performance of the model if left untreated. In this sense, we explore several preprocessing strategies to reduce their impact and obtain a more accurate method. Our best model, a ResNet50 fine-tuned with preprocessed images, obtained 97.84% accuracy when prompted with a single image and 99.50% when processing multiple images from the same patient. In addition to achieving high accuracy, interpretability experiments show that the network correctly learned features from the lung and chest area.

Keywords: COVID-19 diagnostic · CT image analysis · Deep Learning

1 Introduction

As of July 2021—one and a half years after its initial reporting—the COVID-19 (SARS-CoV-2) pandemic has caused more than four million deaths worldwide[1]. From lockdown measures to vaccine research, the outbreak has shaped how governments will approach future health crises to minimize impact in healthcare systems, economic activity and citizen lives. Among most effective measures, mass diagnostic testing was essential to control the spread of the virus [15]. The gold standard diagnostic method for COVID-19 is the Reverse Transcription Polymerase Chain Reaction (RT-PCR). Despite its wide acceptance and high accuracy, its application is laborious, time-consuming, and expensive [14], which hinders mass testing of the population. In this sense, the scientific community urged to research affordable and efficient diagnostic methods for the disease.

As a worldwide event generating a massive amount of data—e.g., patient health records, diagnostic results, social media repercussion—the community

---

[1] https://www.worldometers.info/coronavirus/.

© Springer Nature Switzerland AG 2021
P. F. Stadler et al. (Eds.): BSB 2021, LNBI 13063, pp. 69–80, 2021.
https://doi.org/10.1007/978-3-030-91814-9_7

naturally turned its attention to artificial intelligence [13]. Aiming to leverage the capabilities of data-driven models—i.e., machine learning algorithms able to learn, directly from the input data, the most discriminative features for a given task—researchers proposed deep learning-based diagnostic methods for COVID-19. The approaches act on different types of medical data, such as blood cell counts [4], chest X-rays [10], and computed tomography (CT) scans [2,9,16]. In this work, we are particularly interested in the latter; chest CT scans present a higher sensitivity than X-rays for this problem [3] and have been successfully used in computer-aided diagnosis, such as the classification of respiratory diseases [1].

The recent literature around CT image analysis for COVID-19 diagnosis explores convolutional neural networks (CNN) to classify whether a scan belongs to a healthy or infected patient. CNNs are a type of neural network specialized in processing images, that achieved impressive results in medical imaging analysis [8]. In the case of COVID-19, several researchers work directly on 2D CT scans [2,6,9,16], often fusing model decisions over individual CT slices into a combined diagnosis. Ardakani et al. [2] evaluate several CNN architectures for this problem. Each network is optimized with radiologist-appointed patches from CT slices, and the authors report results that outperform the analyses made by radiologists. Overcoming the need of an expert selecting potential CT regions to be analysed, Xu et al. [16] employ a 3D segmentation CNN to extract regions of interest. The slices are individually classified by a ResNet architecture [7] and the answers are combined to determine the diagnosis. Even though both works achieved interesting results, the datasets used are very limited in size and patient representativity, consisting of less than 1,000 CT samples originated from at most 110 COVID-positive patients. Exploring more representative datasets, Mei et al. [9] combine the analysis of a ResNet over the CT scans with a support vector machine trained over non-image clinical information, such as patient's gender, age, and white cell count. They report promising results in a dataset collected from 905 patients, from which 419 tested positive for SARS-CoV-2.

With newer, bigger and more representative datasets becoming publicly available, more effort is required to curate such data. The quality of a CT sample depends on many aspects, such as the conditions of the sensor, how the subject poses in the patient bed and which clothes they are wearing. When proper data sanitization is not done, low-quality CT scans might be present in the dataset, which might impact machine-aided diagnosis.

In this work, we evaluate several preprocessing techniques to reduce the impact of low-quality CT scans (Fig. 1) into deep learning-based diagnostic tools for COVID-19. We perform our experiments on COVIDx-CT dataset [6], comprised of $104,009$ CT images from $1,489$ patients. We employ both a lightweight baseline CNN as well as a ResNet architecture and report results at image- and exam-level. Finally, we use interpretability techniques [12] to identify which regions of the CT slice have high importance in the model's decision. This acts as a confidence check—to assess if the network is taking into consideration relevant parts of the scan—, but also may guide medical staff into better understanding the disease and how it manifests in patients.

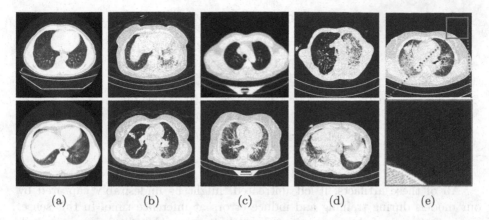

(a)   (b)   (c)   (d)   (e)

Fig. 1. Examples of artifacts present in CT images of the COVIDx-CT dataset [6]: **(a)** rounded borders, **(b)** traces of clothes around chest area, **(c)** structure of the patient bed in the bottom part of the images, **(d)** reflection and bright artifacts outside chest area, and **(e)** background noise.

The remaining of the text is organized as follows. In Sect. 2, we describe COVIDx-CT dataset, presenting the distribution of images per class and examples of existing image artifacts. We present details of our methodology in Sect. 3, describing the overall pipeline of our evaluation and the preprocessing techniques investigated. In Sect. 4, we present the experimental evaluation of all techniques and CNN architectures for this problem. Finally, in Sect. 5, we discuss our final thoughts and draw possible research lines for future work.

2 Dataset

The dataset used in this work is the COVIDx-CT dataset [6]. This database was collected from several hospitals across China and comprises 4, 178 exams of 1, 489 patients, totaling 104, 009 CT images. Each exam is composed of several CT slices and is categorized into three classes: *Normal, Common Pneumonia* (CP), and *Novel Coronavirus Pneumonia* (NCP). Each class was organized by the authors into train, validation and test splits. We present in Table 1 the number of images and patients for each of them.

According to the authors, radiologists manually labeled all scans contained in validation and test sets. Training images, on the other hand, were annotated by non-radiologists, which might yield less trustworthy labels and add to the noisiness of the dataset. Additionally, as the images were captured in different hospitals with distinct equipments, they present a wide variety of quality, resolution, and noisy patterns. Besides sensor-specific noises, they also present artifacts originated from bad subject positioning in the patient bed, traces of the subject's clothes and reflections. Figure 1 illustrates some examples of clear artifacts and noisy patterns in CT scans.

Table 1. Image and patient distributions per set of the COVIDx-CT dataset [6].

	Images			Patients		
	Train	Val	Test	Train	Val	Test
Normal	27,201	9,107	9,450	144	47	52
CP	22,061	7,400	7,395	420	190	125
NCP	12,520	4,529	4,346	300	95	116
Total	61,782	21,036	21,191	864	332	293

All of these artifacts, if left untreated, might be undesirably captured by our models during training and induce errors at inference time. In this sense, we present in the following sections the techniques employed to preprocess each scan and reduce impact on model performance.

3   Methodology

Ultimately, our goal is to accurately classify if a chest CT scan belongs to a patient with COVID-19 or not. Even though we aim to provide a diagnosis at image level—i.e., with a model that processes a single slice and outputs a classification decision among *Normal, CP* and *NCP* for that image—the dataset also groups multiple CT slices of a patient into an exam. This allows us to combine several image-level decisions into an exam-level diagnosis. As each slice captures a different part of the chest and lungs, the rationale is that fusing answers over distinct views might lead to a more reliable and accurate diagnosis.

In our evaluation, we follow the pipeline depicted in Fig. 2, in which each input CT image is preprocessed, as to reduce the impact of capturing and sensor artifacts, and then analyzed by a classification model. Once every image has been classified, their answers are combined for an exam-level diagnosis. In the next sections, we detail the preprocessing techniques, as well as present the CNN architectures used in our evaluation.

3.1   Data Preprocessing

As the region of interest for assessing if a patient has signs of COVID-19 involves mostly the lung areas within the CT image, the preprocessing steps aim to remove the exterior region of the patient's body. Besides not contributing for the diagnosis, this region accounts for most artifacts and noisy patterns seen in Fig. 1. To do so, we extract two masks of the patient's chest region from the original image following the steps presented in Fig. 3.

Initially, we apply 5×5 minimum filters to the original image, suppressing most noisy patterns in the exterior region of the chest. This is followed by a binarization and an edge detection, that yields the contour of the exterior chest area and lungs of the patient. We apply flood fill with seed points in the corners of

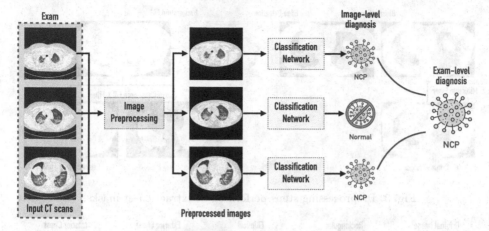

Fig. 2. Overview of our pipeline. An input CT scan is preprocessed to remove possible artifacts and is fed to a CNN that classifies it as *Normal, Common Pneumonia* (CP) or *Novel Coronavirus Pneumonia* (NCP). When an exam is considered, each of its images are processed through the pipeline and the individual image-level diagnosis are combined for an exam-level decision.

the image to remove the background related to the regions outside the patient's body. Finally, the resulting image generates a pair of masks. A subsequent flood fill is performed to obtain the interior chest mask related to the lungs, whereas the exterior chest mask is obtained by applying a sequence of minimum filtering to erase the silhouette of the lungs. With both masks extracted, we explored different ways to apply them to the original image, as depicted in Fig. 4:

– **Rectangular**: the original image is masked with a rectangular region corresponding to the bounding box of exterior chest mask;
– **Rectangular-Centered**: similar to the *Rectangular* approach, but the masked region is centered in the output image;
– **Elliptical**: the original image is masked with a elliptical region that fits the bounding box of exterior chest mask;
– **Elliptical-Centered**: similar to the *Elliptical* approach, but the masked region is centered in the output image;
– **Exterior Chest**: the original image is masked with the exterior chest mask;
– **Interior Chest**: the original image is masked with the interior chest mask and white pixels as background;

All available images in the dataset have equal-sized width and height, ranging from 512×512 to 1024×1024. As they are input to CNNs, which expect fixed-sized images, we avoid performing crops that alter the image's aspect ratio, instead replacing the masked regions with black or white pixels. Besides the masking operation, each image is resized to the CNN's expected input dimensions and pixel values are normalized to the range of $[-1, 1]$.

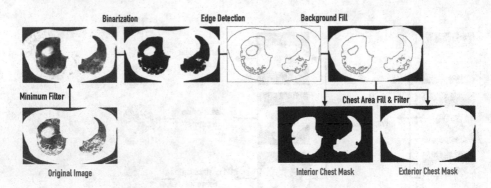

Fig. 3. Preprocessing steps performed to extract chest masks.

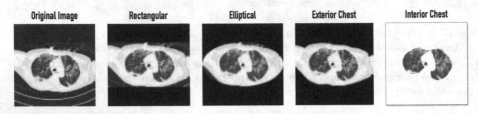

Fig. 4. Original image and examples of generated masked images.

3.2 Deep Neural Network Architectures

In this work, we explore two CNN architectures, a lightweight baseline network trained from scratch and a ResNet50 [7] pretrained on the object classification task of ImageNet dataset [5]. Our goal with the former is to evaluate the impact of each preprocessing technique on a model with lower complexity and optimized solely with the COVIDx-CT dataset. Whereas in our investigations with ResNet50, we aim to fully leverage from a powerful network, employing Transfer Learning [11] to benefit from the knowledge learned in a different task to better generalize in this problem.

The baseline architecture consists of three convolutional blocks, followed by a block of fully-connected layers. Each convolutional block has two convolutional layers, followed by a max pooling with 2×2 kernel and dropout operation with rate 0.2. The number of convolutional filters starts at 32 for the first block and is multiplied by 2 in each subsequent block. The fully-connected block has three fully-connected layers with 256, 128 and 3 neurons, respectively. We employ Scaled Exponential Linear Unit (SELU) as activation function in all layers, except the last fully-connected layer which has a softmax activation.

To adapt the pretrained ResNet for our problem, we performed a series of modifications in its architecture. We remove its last fully-connected layer—which was specific to classify the 1,000 classes of ImageNet—, exchanging it for a global average pooling and two fully-connected layers with 256 and 3 neurons, respectively. Training a parameter-heavy network, such as ResNet, on a dataset with the scale of COVIDx-CT often leads to overfitting. Considering this, we employ dropout operation with rate 0.2 between the last two fully-connected layers. To further minimize the risk of overfitting, we keep the weights of initial layers of the network fixed, only updating the final layers during the training process. In Sect. 4.2, we evaluate the impact of fixing different layers on the performance of COVID-19 diagnosis.

4   Experimental Evaluation

In this section, we present and discuss the experimental evaluation of the preprocessing strategies and CNN architectures considered in this work. Initially, we investigate the impact of each strategy on our baseline architecture. For the best two approaches, we train a ResNet50 varying the number of layers with fixed weights. Considering the best configuration in the previous experiment, we investigate how different image resolutions influence the performance on COVID-19 diagnosis. Finally, we compare the results obtained by our model both at image- and exam-level on the test set of COVIDx-CT dataset, employing explainability techniques to interpret the decisions of the method.

All networks are trained for 20 epochs with Adam optimizer, with a starting learning rate of 10^{-4} and batch size of 32 images of dimensions 128×128. To avoid the negative impact of class unbalance, we use class weights related to the ratio between samples of each diagnosis. To evaluate our approaches, we consider the balanced accuracy and the number of false negatives for the COVID-19 class with respect to the validation set. In this scenario, wrongly classifying a COVID-positive subject as healthy or with common pneumonia might be fatal for the patient and hinders the containment of the virus in our society. Thus, in our analyses, besides achieving high accuracy, we aim to minimize the false negatives for this diagnosis.

4.1   Data Preprocessing

Each preprocessing strategy generates a version of a CT scan with different properties. While *Rectangular* and *Rectangular-Centered* approaches retain some details on the exterior border of the patient's body, the *Elliptical* and *Elliptical-Centered* feed the network with images that fit the shape of the chest more closely while including less of those exterior artifacts. On the other hand, *Exterior Chest* masked images preserve most of the information inside the patient's chest area, whereas *Interior Chest* focuses only on the region containing the lungs.

In this experiment, we evaluate each preprocessing strategy with the baseline architecture. Each model is trained with images transformed by a single approach and we compare their performances with a model trained on the original images, presenting their results in Table 2.

As we expected, image preprocessing greatly impacts performance in this problem. The most accurate model was the one trained with *Interior Chest* strategy. This approach is able to correctly isolate most of the lung region,

Table 2. Balanced accuracy and COVID-19 false negatives with respect to the validation set, considering baseline models trained with each preprocessing strategy. Best result is highlighted in **bold** and second best is underlined.

Preprocessing	↑ Acc (%)	↓ COVID-19 FN (%)
No preprocessing	<u>94.51</u>	2.39
Rectangular	90.93	3.49
Rectangular-centered	94.28	2.63
Elliptical	85.67	6.64
Elliptical-centered	92.42	2.61
Exterior chest	93.39	<u>2.04</u>
Interior chest	**94.53**	**1.31**

allowing the network to focus on the most discriminative area of the CT scan for diagnosing. Surprisingly, the method that achieved the second top accuracy was the one trained with the original images. Even though this seems to be a good result, a model trained with images that retain the artifacts in the exterior of the patient's body might be using such elements as shortcuts to identify the class of an image, instead of learning features within the chest and lung regions. In Sect. 4.4, we perform interpretability experiments to assess this behavior in our models. Finally, when we consider the number of patients with COVID-19 that were diagnosed as non-COVID-19, both *Exterior* and *Interior Chest* strategies obtained the best results. In this sense, we choose this two preprocessing techniques to make further investigations with the ResNet50 architecture.

4.2   Transfer Learning with ResNet50

With *Exterior* and *Interior Chest* strategies selected, we evaluate the ResNet50 architecture. Training such a complex network with a dataset with thousands of images can lead to overfitting. In this sense, we freeze some of its initial layers, fixing their weights learned on the ImageNet pretrain and only updating the weights of the unfrozen layers. We analyze the set of $\{0, 12, 24, 32, 40\}$ unfrozen layers, presenting in Table 3 the results obtained by each network configuration.

In terms of accuracy, the setup with 32 and 40 unfrozen layers and the *Exterior Chest* preprocessing strategy presented a slight advantage over the other models. Considering COVID-19 false negatives, the best configuration— i.e., 32 unfrozen layers with *Exterior Chest* preprocessing strategy—achieved 0.87%, with a balanced accuracy of 96.00%. This setup surpassed the best result obtained by our baseline architecture, which reached 1.31% of COVID-19 false negatives and 94.53% accuracy. Employing a more complex network, while also allowing more layers to be updated improves the network's ability to adapt the previously learned features for this new problem. These results also indicate that focusing only on the lung region (*Interior Chest*) might lose some relevant information that *Exterior Chest* models are able to capture.

Table 3. Balanced accuracy and COVID-19 false negatives with respect to the validation set, considering the ResNet50 architecture trained with different amount of unfrozen layers for the *Exterior Chest* and *Interior Chest* preprocessing strategies. Best result for each metric is highlighted in **bold**.

Unfrozen layers	Exterior chest		Interior chest	
	↑ Acc (%)	↓ COVID-19 FN (%)	↑ Acc (%)	↓ COVID-19 FN (%)
0	93.83	2.58	93.66	2.55
12	94.57	2.57	94.37	2.54
24	95.20	1.85	94.85	2.49
32	96.00	**0.87**	**95.39**	**2.30**
40	**96.21**	1.72	95.13	2.53

Table 4. Evaluation of ResNet50 models, trained with 128×128 and 224×224 resolution images preprocessed by the *Exterior Chest* strategy, with 32 unfrozen layers. Best result is highlighted in **bold**.

Resolution	↑ Acc (%)	↓ COVID-19 FN (%)
128×128	96.00	0.87
224×224	**97.65**	**0.78**

In all previous experiments, the models were trained with images of 128×128 dimensions. Even though available images have higher resolution, the reduced size allowed us to explore several setups and hyperparameter choices. Nonetheless, higher resolution images might present important details that are suppressed when downscaled. In this sense, we investigated the impact of higher resolution on the best configuration of ResNet50 found previously. Table 4 presents the results of the CT images with 128×128 resolution and with 224×224, the default image size for the ResNet50 architecture. The best results were achieved by the model trained with 224×244 images, which obtained 0.78% of COVID-19 false negatives and a balanced accuracy of 97.65%.

4.3 Image- and Exam-Level Classification of the Test Set

Driven by our results on the validation, we evaluated the method on the test set. We considered images individually and also an exam-level classification—i.e. combining the answers for several CT scans of the same patient. We employed the ResNet50 model with 32 unfrozen layers trained on images of 224×224 resolution that were preprocessed by the *Exterior Chest* technique. For the exam-level prediction, we made an ensemble considering a majority voting with the predictions of all the images contained on the exam. Figure 5 presents the confusion matrix of our method for image- and exam-level scenarios on the test set.

The model achieved a high balanced accuracy on both cases, with 97.84% and 99.50% on image- and exam-level, respectively. Nonetheless, combining different

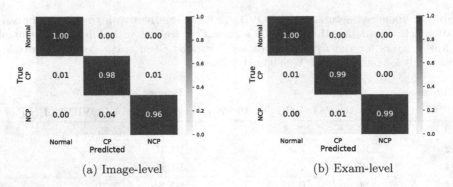

(a) Image-level   (b) Exam-level

Fig. 5. Confusion matrix for image- and exam-levels on the test set.

views over the patient's chest proved to be an essential step for this problem. Besides that, this is more similar to a real application scenario, as each CT exam generates several images from different slices of the lungs. Finally, we also investigated the number of false negatives on COVID-19 cases. On the image-level classification, our model misclassified 0.85% of COVID-19 examples in this set, while in the exam-level classification, only 0.34% of samples were incorrectly assigned as non-COVID.

4.4   Interpreting Model Decisions

Interpretability is an essential characteristic in sensitive tasks, such as medical diagnosis. Being able to understand what made a machine-learning model determine that an image or patient's record has a particular disease is crucial not only to guarantee the correctness of that diagnosis, but also to assess the trustworthiness of the model. It is not enough to have an accurate network, we need to assert it is achieving the correct answers for the right reasons.

Considering this, we use Grad-CAM [12] to generate class-activation maps with the best model of our evaluation. These maps highlight regions of the input image that had a high impact in the model's decision. We present in Fig. 6 maps for a few images of each class, considering models trained with the original images without any preprocessing and *Exterior Chest* masking strategy.

For the *Normal* class, both preprocessing approaches show similar activations, with maps highlighting the central area of the scan. Considering COVID-19 (*NCP*) and Pneumonia (*CP*) diagnoses, images that were not preprocessed often present vestigial background artifacts on the corners of the scans that are highlighted in the activation maps. This indicates that the model learned to identify these artifacts and correlate them to the class label. However, when we consider *Exterior Chest* masked images, this behavior does not happen. The model correctly activates for areas within the patient's lungs, without false activations on the remaining regions of the image.

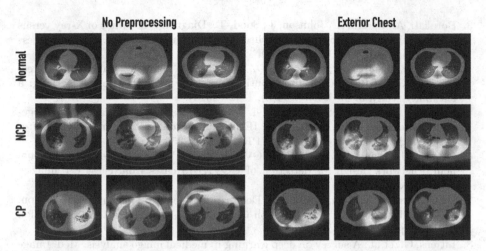

Fig. 6. Class-activation maps [12] for *Normal*, *NCP* and *CP* images from models without any preprocessing and optimized with *Exterior Chest* masking strategy.

5 Conclusion

In this work, we evaluated image preprocessing techniques applied to deep learning-based diagnosis of COVID-19 from CT scans. We trained a lightweight baseline CNN and a ResNet50 on the COVIDx-CT dataset [6]. Scans capture the chest area, but also present artifacts on the exterior region of the patient's body that are originated from bad posing and sensor quality. Our experiments showed the importance of removing this noisy information, as it can be erroneously captured by models and used as a shortcut to infer class label, instead of learning discriminative features for the task. Besides that, carefully increasing model complexity and employing high-resolution images considerably improved the performance in this task. Our best model—a fine-tuned ResNet50 optimized with high-resolution images preprocessed to retain only the chest region of the CT scan—achieved a balanced accuracy of 97.84% with a single image and 99.50% when combining the answers of multiple scans of the same patient. As future work, an ensemble of models could be explored to further improve the results, investigating additional CNN architectures and assessing whether the preprocessing strategies evaluated are complementary with one another.

References

1. Anthimopoulos, M., Christodoulidis, S., Ebner, L., Christe, A., Mougiakakou, S.: Lung pattern classification for interstitial lung diseases using a deep convolutional neural network. IEEE Trans. Med. Imaging **35**(5), 1207–1216 (2016)
2. Ardakani, A.A., Kanafi, A.R., Acharya, U.R., Khadem, N., Mohammadi, A.: Application of deep learning technique to manage COVID-19 in routine clinical practice using CT images: Results of 10 convolutional neural networks. Comput. Biol. Med. **121**, 103795 (2020)

3. Borakati, A., Perera, A., Johnson, J., Sood, T.: Diagnostic accuracy of X-ray versus CT in COVID-19: a propensity-matched database study. Br. Med. J. Open Access (BMJ Open) **10**(11), e042946 (2020)
4. Brinati, D., Campagner, A., Ferrari, D., Locatelli, M., Banfi, G., Cabitza, F.: Detection of COVID-19 infection from routine blood exams with machine learning: a feasibility study. J. Med. Syst. **44**(8), 1–12 (2020)
5. Deng, J., Dong, W., Socher, R., Li, L.J., Li, K., Fei-Fei, L.: ImageNet: a large-scale hierarchical image database. In: IEEE International Conference on Computer Vision and Pattern Recognition (CVPR), pp. 248–255 (2009)
6. Gunraj, H., Wang, L., Wong, A.: COVIDNet-CT: A tailored deep convolutional neural network design for detection of COVID-19 cases from chest CT images. Front. Med. **7** (2020)
7. He, K., Zhang, X., Ren, S., Sun, J.: Deep residual learning for image recognition. In: IEEE International Conference on Computer Vision and Pattern Recognition (CVPR), pp. 770–778 (2016)
8. Litjens, G., et al.: A survey on deep learning in medical image analysis. Med. Image Anal. **42**, 60–88 (2017)
9. Mei, X., et al.: Artificial intelligence-enabled rapid diagnosis of patients with COVID-19. Nat. Med. **26**(8), 1224–1228 (2020)
10. Oliveira, G., et al.: COVID-19 X-ray image diagnostic with deep neural networks. In: BSB 2020. LNCS, vol. 12558, pp. 57–68. Springer, Cham (2020). https://doi.org/10.1007/978-3-030-65775-8_6
11. Raghu, M., Zhang, C., Kleinberg, J., Bengio, S.: Transfusion: understanding transfer learning for medical imaging. In: Advances in Neural Information Processing Systems (NIPS), pp. 3347–3357 (2019)
12. Selvaraju, R.R., Cogswell, M., Das, A., Vedantam, R., Parikh, D., Batra, D.: Gradcam: Visual explanations from deep networks via gradient-based localization. In: IEEE International Conference on Computer Vision (ICCV), pp. 618–626 (2017)
13. Shi, F., et al.: Review of artificial intelligence techniques in imaging data acquisition, segmentation, and diagnosis for COVID-19. IEEE Rev. Biomed. Eng. **14**, 4–15 (2020)
14. Smyrlaki, I., et al.: Massive and rapid COVID-19 testing is feasible by extraction-free SARS-CoV-2 RT-PCR. Nat. Commun. **11**(1), 1–12 (2020)
15. Vandenberg, O., Martiny, D., Rochas, O., van Belkum, A., Kozlakidis, Z.: Considerations for diagnostic COVID-19 tests. Nat. Rev. Microbiol. **19**(3), 171–183 (2021)
16. Xu, X., et al.: A deep learning system to screen novel coronavirus disease 2019 pneumonia. Engineering **6**(10), 1122–1129 (2020)

Feature Importance Analysis of Non-coding DNA/RNA Sequences Based on Machine Learning Approaches

Breno Lívio Silva de Almeida[1], Alvaro Pedroso Queiroz[2],
Anderson Paulo Avila Santos[1,3](✉), Robson Parmezan Bonidia[1],
Ulisses Nunes da Rocha[3], Danilo Sipoli Sanches[2],
and André Carlos Ponce de Leon Ferreira de Carvalho[1]

[1] Institute of Mathematics and Computer Science, University of São Paulo - USP,
São Carlos 13566-590, Brazil
anderson.avila@usp.br
[2] UTFPR, Federal University of Technology-Paraná, Campus Cornélio Procópio,
Cornélio Procópio 86300-000, Brazil
[3] Department of Environmental Microbiology, Helmholtz Centre for Environmental
Research – UFZ GmbH, 04318 Leipzig, Saxony, Germany

Abstract. Non-coding sequences have been gained increasing space in
scientific areas related to bioinformatics, due to essential roles played
in different biological processes. Elucidating the function of these non-
coding regions is a relevant challenge, which has been addressed by several Machine Learning (ML) studies in various fields of ncRNA, e.g.,
small non-coding RNAs (sRNAs) and Circular RNAs (circRNAs). The
identification of these biological sequences is possible through feature
engineering techniques, which can help point out specifics in different
types of problems with ML. Thereby, there are recent studies focusing on
interpretable computational methods, i.e., the best features based on feature importance analysis. For that reason, in this study we have proposed
to explore different features descriptors and the degree of importance
involved for classification task, using two case studies: (1) prediction of
sRNAs in Bacteria and (2) prediction of circRNA in Humans. We developed a general pipeline using hybrid feature vectors with mathematical
and conventional descriptors. In addition, these vectors were generated
with MathFeature package and feature selection techniques in both case
studies. Finally, our experiments results reported high predictive performance and the relevance of combining conventional and mathematical
descriptors in different organisms.

Keywords: Machine learning · Small RNA · Feature extraction ·
Feature importance · MathFeature

B. L. S. de Almeida, A. P. Queiroz and A. P. A. Santos—The authors wish it to be
known that, in their opinion, the first three authors should be regarded as Joint First
Authors.

P. F. Stadler et al. (Eds.): BSB 2021, LNBI 13063, pp. 81–92, 2021.
https://doi.org/10.1007/978-3-030-91814-9_8

1 Background

Over the years, with the accelerated advance of biological studies, non-coding RNAs (ncRNA) sequences have attracted attention in bioinformatics [12]. According to [34], recent studies have shown that ncRNAs can play essential roles in biological processes, e.g., transcriptional regulation [24], epigenetics [26], and human diseases [3]. The identification of ncRNA sequences is fundamental to better understand their mechanisms and functions, but elucidating the function of these non-coding regions is a key challenge [31], which has been addressed by several ML studies.

Remarkable studies of ML have shown relevant results in various fields of ncRNA, e.g., bacterial sRNAs, which are currently being discovered as important elements in various physiological processes, including growth, development, cell proliferation, differentiation, and metabolic reactions [28,29]; and circRNAs, which have been identified in many species, including humans, mouse, plants, and archaea [14,21].

Nevertheless, ML studies applied to biological data require the extraction of features to identify patterns that allow their classification among other sequences. ML models need to understand in a relatively concrete way the characteristics of biological sequences, such as their statistics and other relevant information. Thereby, the feature extraction plays an important step in ML [9]. Moreover, biological data has a higher dimensional nature, generating a large number of variables, where feature selection and dimensionality reduction techniques must be considered in conjunction with the feature extraction [33].

Techniques such as Deep Learning can do the process of feature extraction by themselves, but problem-specific knowledge needs to be carefully modeled to address, e.g., biological sequences and text mining problems [35]. Cases in which Deep Learning struggles with a reasonable representation, considering known feature engineering methods can help with the model, as we can see in [22], which uses k-mer representation. Nevertheless, feature engineering yet can be a crucial step to better understand the data and its relevant characteristics.

In scientific areas related to bioinformatics, finding the most significant feature (feature importance) generates relevant contributions to the interpretability of the model [32], allowing the understanding of the internal decision-making process [8]. Studies have focused on this theme, e.g., [8] proposed a model based on decomposing solutions for Long non-coding RNAs (lncRNAs), in which the least relevant features are suppressed according to their contribution to a classification task. [2] reported an exploratory data analysis to select the most important features for the ncRNA identification. [13] explored features that could distinguish cirRNAs from other lncRNAs. These studies have in common the search for better features and the use of conventional feature descriptors, e.g., (Open Reading Frame (ORF) and k-mer), following the pattern of several other approaches related to ncRNAs [9].

Therefore, there is an increasing demand for interpretable computational methods, involving the understanding of the best features [11]. In this study, we propose a pipeline using hybrid feature vectors (mathematical (little covered in ncRNA [9]) and conventional descriptors - to generate this hybrid vector, we use the MathFeature package [10]) with feature selection techniques and dimensionality reduction in two case studies, e.g., (1) prediction of sRNAs in Bacteria and (2) prediction of circRNA in Humans. Thereby, we assume the following problematic:

– **Problematic:** Based on a hybrid feature vector with mathematical and conventional descriptors, which are the most important features for classifying sRNA (Bacteria) and circRNA (human) sequences?

Our experimental results reported two contributions: (1) High predictive performance in both case studies and (2) the relevance of using the hybrid combination of mathematical and conventional descriptors.

2   Pipeline for Machine Learning Classification Task

As aforementioned, we investigated different feature descriptors (descriptor refers to the feature extraction technique that can present several values/measures) applied in two domain problems, such as: prediction of sRNAs in Bacteria and prediction of circRNA in Humans. For that reason, we proposed a general pipeline considering our benchmarks for predictive task analysis as illustrated in Fig. 1. Moreover, we evaluated a vector of hybrid features (mathematical and conventional) and different ML models behave in distinct organisms, e.g., prokaryotes and eukaryotes. More details regarding to proposed pipeline are introduced in the next subsections.

2.1   Prediction of sRNAs

Firstly, we described the pipeline for classification task of sRNAs in bacteria, where we have selected curated sRNA sequences from [5], which also proposed an improved method for classifying sRNA sequences. The data used are experimentally-validated sRNAs of Salmonella Typhimurium LT2 (SLT2), collected by [4] using RFAM database [18]. Authors in [5] created negative datasets using EMBOSS' shuffleseq [25] and it was also used for reproducibility and better comparison. In total, we have used 182 sRNA vs. 182 non-sRNA (shuffled) sequences in our pipeline. In addition, we have applied MathFeature framework, a feature extraction package, that provides 37 descriptors based on several studies found in the literature, e.g., multiple numeric mappings, genomic signal processing, entropy, and complex networks [10]. As seen in [9], approaches using graphs and entropy seem suitable methods to use with ML models. The package also has conventional descriptors as Dinucleotide Composition (DNC) and Trinucleotide Composition (TNC) used by [5], who reported a reasonable classification.

For prediction of sRNAs, we used the following feature descriptors: Complex Networks (Graphs with $k = 1, 2, \ldots, 6$ (13 features each k)), Shannon and Tsallis Entropy, ORF, Fickett Score, DNC, and TNC. We also assess the degree of

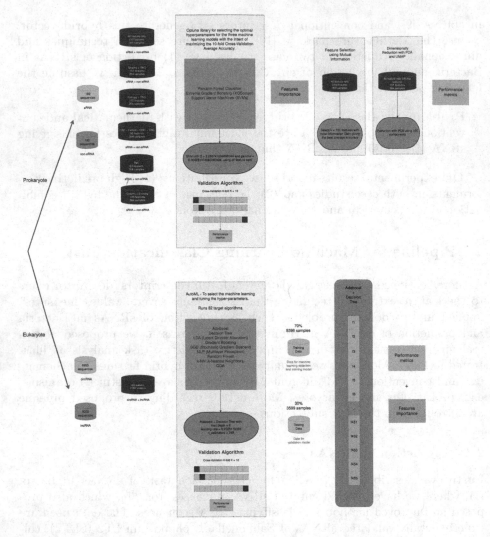

Fig. 1. Pipeline for prediction task of sRNA and circRNA

influence of each feature, applying feature selection techniques available in the literature [7]. In that case, it is desirable to reduce the number of input variables to reduce both the computational cost of modeling and, in some cases, to improve the performance of the model. We have applied Mutual Information (MI) which is a powerful method for detecting relationships between data sets [27], in which variables with the most Information Gain were selected. To assess our model, we use Accuracy (ACC), precision, F1-score, and Area Under the Curve (AUC).

Hyperparameter Optimization and Dimensionality Reduction: We induced Support Vector Machines (SVMs) for classification (Scikit-learn [23] library), whereas this classifier had the best performance in [5]. For comparison, we also induced the Random Forest (RF) and XGBoost [15] classifiers. The Optuna library [1] was used to find the most adequate parameters for the classification models. Optuna uses the TPE (Tree-structured Parzen Estimator), which is a Bayesian optimization algorithm [1]. Optuna framework can be faster and more efficient than popular methods such as Grid Search or Random Search [16]. Thereby, ten thousand trials were performed with Optuna for each case, fixing the SVM kernel as RBF but varying C (regularization parameter) and γ (kernel coefficient). Finally, we have also included dimensionality reduction analysis, which attempts a lower-dimensional representation of the numerical input data that preserves prominent relationships in the data. The dimensionality reduction analysis is known for being used with linear techniques such as Principal Component Analysis (PCA), but they cannot handle complex nonlinear data properly [30]. Considering this, we have applied recent state-of-art techniques as Uniform Manifold Approximation and Projection (UMAP) [20] for representing and visualizing the data. As seen in [6], non-linear dimensionality reduction techniques are being more recognized and can avoid overcrowding of the representation, wherein different clusters are represented on an overlapping area.

2.2 Prediction of CircRNA

In our second study, we used the benchmark dataset provided by [9] for the prediction of circRNA versus lncRNA with 11,995 sequences (6,995 circRNA and 5,000 lncRNA). In addition, we also used the MathFeature framework [9], extracting feature descriptors such as Nucleic acid composition (NAC), DNC, TNC, Xmer k-Spaced Ymer Composition Frequency (kGap), Pseudo K-tuple Nucleotide Composition (PseKNC), Fourier transform with complex numbers, Shannon entropy, Fickett score and ORF. We also evaluated the influence degree of each feature in this case study, using Gini importance of features as a reference index to determine the optimal feature subset [19].

Hyperparameter Optimization: After the feature extraction process, we generated a dataset with 455 features and 11,995 samples. We divided the dataset into 70% of the samples (8,396) for training (10-fold cross-validation) and 30% (3,599) for testing. To select the best classifiers and hyperparameters, an Automated Machine Learning technique (AutoML) was used, where all these steps are performed automatically using the efficient Bayesian optimization method [17], assessing 82 combinations of classifiers, with the best performance being the combination of the ensemble learning technique, Adaboost, combined with Decision Tree (maximum depth equal to 6, 245 estimators and learning rate equal to 0.2225). To assess our model, we use ACC, precision, F1-score, and AUC.

3   Results and Discussion

3.1   Case Study 1: sRNAs in Bacteria

In this case study, our experimental results show that mathematical descriptors such as Complex Networks, Shannon, and Tsallis Entropy can contribute to sRNA classification with conventional descriptors like ORF, Fickett Score, DNC, and TNC. We realized that TNC descriptor is a suitable descriptor for the classification as seen in [5], however, achieved a considerable improvement when combining different descriptors. It's important to add that the same shuffled sequences, the first negative dataset in [5], was used for all training and testing, considering how the model performance can be influenced by how these sequences are shuffled. The ratio for training and testing was 1:1 and the results shown are the average 10-fold cross-validation for each metric, as our reference [5]. These results can be seen in Table 1.

Table 1. SVM model performance with optimal parameters.

Feature vector	ACC	Precision	F1-Score	AUC
Top features ·	**0.8928**	0.9205	**0.8890**	0.9398
Full	0.8899	**0.9272**	0.8850	0.9353
Graphs + TNC	0.8875	0.9177	0.8805	**0.9450**
Entropy + TNC	0.8847	0.9098	0.8786	0.9385
ORF + Fickett + DNC + TNC	0.8845	0.9096	0.8805	0.9337
TNC	0.8736	0.8828	0.8719	0.9277
Graphs + Entropy	0.7803	0.8169	0.7663	0.8515

Essentially, the RF and XGBoost classifiers reported lower average performance with 0.8572 and 0.8769, respectively, for average ACC using full feature vector. For SVM, the highest average ACC score was obtained when combining all features with feature selection (MI). Furthermore, features based on complex networks have an important highlight in terms of information gain (see Fig. 2). Nevertheless, feature vectors based on conventional descriptors have all their features used (see Table 2). While mathematical descriptors such as complex networks and Tsallis entropy showed a reasonable proportion of features.

Table 2. Proportion of features used for each feature descriptor.

Feature vector	Dimension	Proportion used (%)
Complex networks	78	48.7179
Shannon entropy	24	12.5000
Tsallis entropy	24	75.0000
ORF	10	100
Fickett score	2	100
DNC	16	100
TNC	64	100

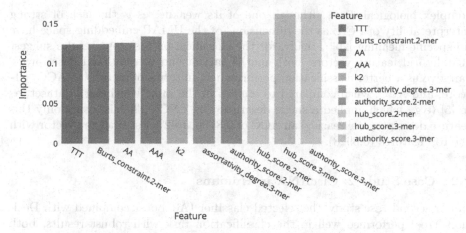

Fig. 2. Top 10 features with most information gain using MI.

In addition, posteriorly to feature importance analysis, we apply PCA for dimensionality reduction, to visualize and represent the data. The dataset was reduced to 100 principal components, maintaining the same classification performance. However, data visualization with PCA was not found to be a good representation of the data, since it showed notable variance beyond the first three components (see Fig. 3 (a)). Thus, for visualization, we adopted UMAP, which can create a better representation, considering the observed non-linearity (see Fig. 3 (b)).

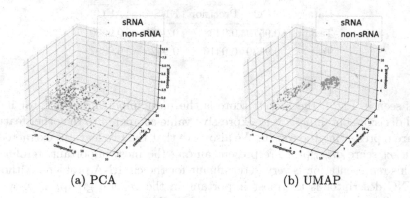

(a) PCA   (b) UMAP

Fig. 3. Representation in three dimensions using dimensionality reduction.

Non-linear dimensionality reduction techniques as UMAP are known to preserve more of the global structure than linear techniques as PCA, and arguably even than other non-linear techniques such as t-distributed Stochastic Neighbor Embedding (t-SNE) [20]. These techniques are essential to represent non-linear complex biological data, although one of its weaknesses is the lack of strong interpretability of PCA, as the dimensions of the UMAP embedding space have no specific meaning [20]. Finally, for the classification task, our results suggest that hybrid feature vectors, both mathematical and conventional descriptors, can give us a better classification performance, in terms of metrics as ACC, F1-score, and AUC, when compared to studies in the same benchmark dataset, as in [5] (e.g., which applied a single descriptor as TNC). That is, using only this feature descriptor, we reached an ACC of 0.8736, 1.92% less than the vector with the top features (0.8928).

3.2   Case Study 2: CircRNA in Humans

In our second case study, the selected classifier (Adaboost combined with Decision Tree) performed well in the classification task with robust results, both in training and testing, as can be seen in Table 3. According to the results, our model reported performance with ACC, precision, F1-score, AUC of 0.9530, 0.9416, 0.9425, and 0.9897, respectively. Moreover, we realized that the feature extraction step contributes positively to the definition of the non-coding RNA classes, with all metrics close to 0.9500. Our pipeline reached a low rate of false positives. Based on this, we investigated which features are most important to classify circRNA and lncRNA. In that case, Fig. 4, provides the best 10 features, calculated through the Gini importance in a normalized way. We also elaborated the Table 4 with the importance score by descriptor.

Table 3. Performance on training and testing dataset.

Dataset	ACC	Precision	F1-Score	AUC
Training	0.9579	0.9498	0.9495	0.9930
Test	0.9530	0.9416	0.9425	0.9897

As seen in Fig. 4, the Fickett score is the most important descriptor in the model decision, obtaining a very expressive value compared to the other features that are representing the top 10. We also note that the two features extracted by the Ficket score descriptor are present among the most important, reinforcing that this representation is very determinant for the circRNA problem. Although the TNC descriptor is the most important in the general group, as shown in Table 4, it generates a large number of features (64), while Fickett score generates only two. Therefore, this information can be investigated by biologists to try to understand the meaning of these combinations of nitrogenous bases, and how they can be used to differentiate circular and long non-coding RNAs.

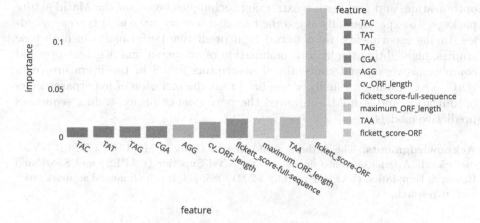

feature
- TAC
- TAT
- TAG
- CGA
- AGG
- cv_ORF_length
- fickett_score-full-sequence
- maximum_ORF_length
- TAA
- fickett_score-ORF

feature

Fig. 4. The best 10 features based on Gini importance (normalized).

Table 4. Feature importance - circRNA versus lncRNA.

Descriptor	Importance score
TNC	0.375193
Ficket score	0.157546
Complex networks	0.117040
ORF	0.081057
MonoDiKGap	0.073297
DiMonoKGap	0.072506
DNC	0.054250
Fourier transform with complex numbers	0.020534
PseKNC - type 2	0.018784
MonoMonoKGAP	0.011886
Shannon entropy	0.009029
NAC	0.008878

Finally, the importance of the features extracted by the TNC descriptor represents about 37.7% of the total score. However, we emphasize that the division of importance between other descriptors, as the Fickett score and Complex Network, are extremely significant, again indicating the relevance of our proposal with hybrid vectors.

4   Conclusion

In this paper, we have proposed a general pipeline to deal with different domains, prediction of sRNA in bacteria and prediction of circRNA in humans. For that reason, we have proposed different ML approaches with efficient hyperparameters optimization, and also feature extraction techniques based on the MathFeature package. To experimentally assess the best features, we have used two case studies. In the experiments, we obtained high predictive performance in both case studies, highlighting the hybrid combination of mathematical (e.g., entropy and complex networks) and conventional descriptors found in the literature (e.g., AAC, TNC, Coding). Finally, we realized that the inclusion of hyperparameters optimization and AutoML improved the performance of non-coding sequences predictive models.

Acknowledgments. The authors would like to thank ICMC-USP, UTFPR, Coordenação de Aperfeiçoamento de Pessoal de Nível Superior (CAPES) and São Paulo Research Foundation (FAPESP), grant #2021/08561-8, for the financial support given to this research.

References

1. Akiba, T., Sano, S., Yanase, T., Ohta, T., Koyama, M.: Optuna: a next-generation hyperparameter optimization framework. In: Proceedings of the 25th ACM SIGKDD International Conference on Knowledge Discovery & Data Mining, pp. 2623–2631 (2019)
2. Amin, N., McGrath, A., Chen, Y.P.P.: Fexrna: exploratory data analysis and feature selection of non-coding rna. IEEE/ACM Trans. Comput. Biol. Bioinform. 1 (2021). https://doi.org/10.1109/TCBB.2021.3057128
3. Anastasiadou, E., Jacob, L.S., Slack, F.J.: Non-coding RNA networks in cancer. Nat. Rev. Canc. **18**(1), 5–18 (2018)
4. Arnedo, J., Romero-Zaliz, R., Zwir, I., Del Val, C.: A multiobjective method for robust identification of bacterial small non-coding RNAs. Bioinformatics **30**(20), 2875–2882 (2014)
5. Barman, R.K., Mukhopadhyay, A., Das, S.: An improved method for identification of small non-coding RNAs in bacteria using support vector machine. Sci. Rep. **7**(1), 1–8 (2017)
6. Becht, E., et al.: Dimensionality reduction for visualizing single-cell data using UMAP. Nat. Biotech. **37**(1), 38–44 (2019)
7. Bommert, A., Sun, X., Bischl, B., Rahnenführer, J., Lang, M.: Benchmark for filter methods for feature selection in high-dimensional classification data. Comput. Stat. Data Anal. **143**, 106839 (2020). https://doi.org/10.1016/j.csda.2019.106839
8. Bonidia, R.P., et al.: A novel decomposing model with evolutionary algorithms for feature selection in long non-coding RNAs. IEEE Access **8**, 181683–181697 (2020). https://doi.org/10.1109/ACCESS.2020.3028039
9. Bonidia, R.P., et al.: Feature extraction approaches for biological sequences: a comparative study of mathematical features. Briefings Bioinform. **22**(5), bbab011 (2021). https://doi.org/10.1093/bib/bbab011
10. Bonidia, R.P., Sanches, D.S., de Carvalho, A.C.: Mathfeature: feature extraction package for biological sequences based on mathematical descriptors. bioRxiv (2020)
11. Carvalho, D.V., Pereira, E.M., Cardoso, J.S.: Machine learning interpretability: a survey on methods and metrics. Electronics **8**(8), 832 (2019)
12. Chantsalnyam, T., Siraj, A., Tayara, H., Chong, K.T.: ncRDense: a novel computational approach for classification of non-coding RNA family by deep learning. Genomics **113**(5), 3030–3038 (2021). https://doi.org/10.1016/j.ygeno.2021.07.004

13. Chen, L., et al.: Discriminating cirRNAs from other lncRNAs using a hierarchical extreme learning machine (H-ELM) algorithm with feature selection. Mol. Gen. Genomics **293**(1), 137–149 (2018)
14. Chen, L., et al.: The bioinformatics toolbox for circRNA discovery and analysis. Briefings Bioinform. **22**(2), 1706–1728 (2020). https://doi.org/10.1093/bib/bbaa001
15. Chen, T., Guestrin, C.: XGBoost: A scalable tree boosting system. In: Proceedings of the 22nd ACM SIGKDD International Conference on Knowledge Discovery and Data Mining, pp. 785–794. KDD 2016, ACM, New York, NY, USA (2016). https://doi.org/10.1145/2939672.2939785
16. Ekundayo, I.: OPTUNA Optimization Based CNN-LSTM Model for Predicting Electric Power Consumption. Ph.D. thesis, Dublin, National College of Ireland (2020)
17. Feurer, M., Klein, A., Eggensperger, K., Springenberg, J., Blum, M., Hutter, F.: Efficient and robust automated machine learning. In: Cortes, C., Lawrence, N., Lee, D., Sugiyama, M., Garnett, R. (eds.) Advances in Neural Information Processing Systems, vol. 28. Curran Associates, Inc. (2015)
18. Griffiths-Jones, S., Moxon, S., Marshall, M., Khanna, A., Eddy, S.R., Bateman, A.: Rfam: annotating non-coding RNAs in complete genomes. Nucleic Acids Res. **33**(suppl_1), D121–D124 (2005)
19. Lin, L., Wang, D., Zhao, S., Chen, L., Huang, N.: Power quality disturbance feature selection and pattern recognition based on image enhancement techniques. IEEE Access **7**, 67889–67904 (2019). https://doi.org/10.1109/ACCESS.2019.2917886
20. McInnes, L., Healy, J., Melville, J.: Umap: Uniform manifold approximation and projection for dimension reduction. arXiv preprint arXiv:1802.03426 (2018)
21. Niu, M., et al.: CirRNAPL: a web server for the identification of circRNA based on extreme learning machine. Comput. Struct. Biotechn. J. **18**, 834–842 (2020)
22. Noviello, T.M.R., Ceccarelli, F., Ceccarelli, M., Cerulo, L.: Deep learning predicts short non-coding RNA functions from only raw sequence data. PLoS Computat. Biol. **16**(11), e1008415 (2020)
23. Pedregosa, F., et al.: Scikit-learn: machine learning in python. J. Mach. Learn. Res. **12**, 2825–2830 (2011)
24. Pisignano, G., Ladomery, M.: Post-transcriptional regulation through long non-coding RNAs (lncRNAs). Non-Coding RNA **7**(2) (2021). https://doi.org/10.3390/ncrna7020029
25. Rice, P., Longden, I., Bleasby, A.: Emboss: the European molecular biology open software suite. Trends Genet. **16**(6), 276–277 (2000)
26. Rong, D., et al.: Epigenetics: roles and therapeutic implications of non-coding RNA modifications in human cancers. Mol. Ther.-Nucleic Acids (2021)
27. Ross, B.C.: Mutual information between discrete and continuous data sets. PLoS One **9**(2), e87357 (2014)
28. Strobel, E.J., Watters, K.E., Loughrey, D., Lucks, J.B.: Rna systems biology: uniting functional discoveries and structural tools to understand global roles of RNAs. Curr. Opin. Biotechnol. **39**, 182–191 (2016). https://doi.org/10.1016/j.copbio.2016.03.019, systems biology • Nanobiotechnology
29. Tang, G., Shi, J., Wu, W., Yue, X., Zhang, W.: Sequence-based bacterial small RNAs prediction using ensemble learning strategies. BMC Bioinf. **19**(20), 13–23 (2018)
30. Van Der Maaten, L., Postma, E., Van den Herik, J., et al.: Dimensionality reduction: a comparative. J. Mach. Learn. Res. **10**(66–71), 13 (2009)

31. Vitsios, D., Dhindsa, R.S., Middleton, L., Gussow, A.B., Petrovski, S.: Prioritizing non-coding regions based on human genomic constraint and sequence context with deep learning. Nat. Commun. **12**(1), 1–14 (2021)
32. Wei, G., Zhao, J., Feng, Y., He, A., Yu, J.: A novel hybrid feature selection method based on dynamic feature importance. Appl. Soft Comput. **93**, 106337 (2020). https://doi.org/10.1016/j.asoc.2020.106337
33. Yamada, M., et al.: Ultra high-dimensional nonlinear feature selection for big biological data. IEEE Trans. Knowl. Data Eng. **30**(7), 1352–1365 (2018)
34. Zhong, L., Zhen, M., Sun, J., Zhao, Q.: Recent advances on the machine learning methods in predicting ncRNA-protein interactions. Mol. Genet. Genomics **296**(2), 243–258 (2021)
35. Zhou, S., Li, X.: Feature engineering vs. deep learning for paper section identification: toward applications in Chinese medical literature. Inf. Process. Manag. **57**(3), 102206 (2020)

Heuristics for Cycle Packing
of Adjacency Graphs for Genomes
with Repeated Genes

Gabriel Siqueira$^{(\boxtimes)}$ (iD), Andre Rodrigues Oliveira (iD),
Alexsandro Oliveira Alexandrino (iD), and Zanoni Dias (iD)

Institute of Computing, University of Campinas (Unicamp), Campinas, Brazil
{gabriel.siqueira,andrero,alexsandro,zanoni}@ic.unicamp.br

Abstract. The Adjacency Graph is a structure used to model genomes
in several rearrangement distance problems. In particular, most studies use properties of a maximum cycle packing of this graph to develop
bounds and algorithms for rearrangement distance problems, such as
the reversal distance and the Double Cut and Join (DCJ) distance.
When each genome has no repeated genes, there exists only one cycle
packing for the graph. However, when each genome may have repeated
genes, the problem of finding a maximum cycle packing for the adjacency
graph (Adjacency Graph Packing) is NP-hard. In this work, we developed a greedy random heuristic and a genetic algorithm heuristic for the
Adjacency Graph Packing problem for genomes with repeated genes.
We present experimental results and compare these heuristics with the
SOAR framework. Furthermore, we show how the solutions from our
algorithms can improve the estimation for the reversal distance when
compared to the SOAR framework. Lastly, we applied our genetic algorithm heuristic in real genomic data to validate its practical use.

Keywords: Genome rearrangements · Reversals · Cycle packing

1 Introduction

In the course of evolution, genomes undergo mutations. These mutations can be
punctual (i.e., insertion, deletion or duplication) [8,15] or large-scale rearrangements, such as *genome rearrangements*, altering a large stretch of a genome.

In Comparative Genomics, a well-accepted way to infer the distance between
two genomes of closely related species is by computing the minimum number of
large-scale mutations needed to transform one genome into the other. Several
genome rearrangements have already been proposed and studied, such as the
reversal [1], where a block of genes from the genome is inverted, and the *Double-
Cut-and-Join* (DCJ) [2], that cuts the genome at two positions and creates two
new adjacencies by joining the four extremities affected by these cuts.

The first works on genome rearrangements considered genomes sharing the
same set of genes, and with no repeated genes. These restrictions allow to repre-

© Springer Nature Switzerland AG 2021
P. F. Stadler et al. (Eds.): BSB 2021, LNBI 13063, pp. 93–105, 2021.
https://doi.org/10.1007/978-3-030-91814-9_9

sent gene order using a permutation, where each element has a '+' or '-' sign indicating gene orientation. The reversal distance and the DCJ distance on signed permutations can be solved in polynomial time [2,7].

Later, models considering duplicated or multiple copies of the same gene started to be studied. In this case, it is common to represent genomes as strings, where genes are represented by labels, and repeated genes use the same label across the string. Similar to the genome representation using permutations, when gene orientation is known it can be represented by signs. Although more realistic, most of these problems belong to the class of NP-hard problems [3,4,12].

Most results for the rearrangement distance use the adjacency graph (or the equivalent structured called breakpoint graph) [2,3,7] to model the two compared genomes. This structure is defined in Sect. 2. More precisely, bounds and algorithms rely on cycle packings of this graph. When there are no repeated genes, there exists only one cycle packing and it is trivial to find it [7].

Any cycle packing of an adjacency graph induces a one-to-one assignment between genes of same label. The size of a cycle packing, meaning the number of cycles in it, is inversely related to the value of upper and lower bounds for the reversal and DCJ distance. In this way, we are interested in finding maximum size cycle packings of adjacency graphs. However, when there are repeated genes, the problem of finding a maximum size cycle packing of an adjacency graph, called *Adjacency Graph Packing* problem, is NP-hard [13]. A similar work on cycle packing was recently presented by Pinheiro *et al.* [11], but they assumed no repeated genes, and they did not consider gene orientation.

In this work, we create heuristics to find cycle packings of an adjacency graph with as many cycles as possible when considering repeated genes and gene orientation. We present two heuristics, one using a randomized greedy strategy, and one using a Genetic Algorithm [10]. We test our heuristics in real and simulated genomes and compare them against the SOAR framework [3].

This work is divided as follows. Section 2 presents all the definitions used throughout the manuscript. Section 3 proposes a randomized greedy heuristic for the Adjacency Graph Packing problem. Section 4 defines a heuristic based on the Genetic Algorithm technique. Section 5 presents the experiments with the proposed heuristics and the SOAR framework. Section 6 concludes the paper.

2   Definitions

A genome $\mathcal{G} = (g_1 \; g_2 \; \ldots \; g_n)$ is treated as a sequence of n genes labeled in order as $g_1, g_2, \ldots, g_n$. We represent $\mathcal{G}$ by a *signed string* S, where each character S_i (i.e., the character at position i) corresponds to a gene g_i and it has a + or − sign representing the gene orientation.

The *size* of S is denoted by $|S|$. The *alphabet* of a string S, denoted by Σ_S, is the set of all characters of S ignoring their signs. In this work, we use numbers as the characters of the strings that represent the genomes.

A character $\alpha \in \Sigma_S$ appears in S one or more times, and each one is called an *occurrence* of α in S. The number of occurrences of α in S is denoted by $occ(\alpha, S)$.

Given a string S, a character α is called a *singleton*, if $occ(\alpha, S) = 1$, and it is called *replicated* otherwise.

Example 1. A string S and some information of this string. The character 2 is a singleton, while all other characters are replicated.

$$S = (+1 \ +2 \ -1 \ +4 \ -3 \ +1 \ -4 \ +3)$$
$$\Sigma_S = \{1, 2, 3, 4\}, |S| = 8, S_1 = +1, S_7 = -4$$
$$occ(1, S) = 3, occ(2, S) = 1, occ(3, S) = occ(4, S) = 2$$

We say that two strings S and P are *balanced* if $\Sigma_S = \Sigma_P$ and $occ(\alpha, S) = occ(\alpha, P)$, for every $\alpha \in \Sigma_S$. Note that balanced strings must have the same size. In this work, we will assume that all strings are balanced.

We *extend* a string S by adding the elements $S_0 = +0$ and $S_{|S|+1} = +(|S|+1)$. Henceforward, we consider that strings are in their extended form.

Given a string S, we define the *partial graph* $G_S = (V_S, E_S)$. For every $0 \leq i \leq |S|$, there exists two vertices in V_S of labels $+S_i$ and $-S_{i+1}$ connected by an undirected edge $e_i^S \in E_S$, which is classified as a *black edge*. Note that each character S_i in S, except for $+0$ and $+(|S|+1)$, corresponds to two vertices $+S_i$ and $-S_i$, and we say that such vertices are *twins*.

Given two balanced strings S and P, the *adjacency graph* $G(S, P) = (V, E)$ includes the disjoint union of the partial graphs G_S and G_P, and the set of gray edges E_g. For every pair $u \in V_S$ and $v \in V_P$, there exists a gray edge $\{u, v\}$ in E_g if these vertices have the same label. Two gray edges $\{u, v\}$ and $\{t, s\}$ are called *twins* if u and t are twins, and v and s are also twins. Figure 1 shows an example of the adjacency graph for two balanced strings.

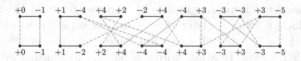

Fig. 1. The adjacency graph of two balanced strings $S = (+1 + 4 - 2 - 4 - 3 + 3)$ and $P = (+1 + 2 - 4 + 4 - 3 - 3)$. Black edges are represented by continuous lines and gray edges are represented by dashed lines. Some twin gray edges are represented by dotted lines.

An *alternating cycle* of an adjacency graph $G(S, P)$ is a cycle composed of alternating edges between blacks and grays. An *alternating cycle packing* (or just *cycle packing*) of the graph $G(S, P)$ is a set of disjoint alternating cycles, such that: (i) each vertex belongs to exactly one alternating cycle; (ii) a gray edge $\{u, v\}$ belongs to a cycle iff it does not have a twin or its twin edge is also in some cycle. Figure 2 shows a cycle packing of the adjacency graph from Fig. 1.

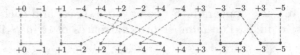

Fig. 2. A cycle packing for the adjacency graph of two balanced strings $S = (+1 + 4 - 2 - 4 - 3 + 3)$ and $P = (+1 + 2 - 4 + 4 - 3 - 3)$. The cycle packing has three different alternating cycles.

Given an alternating cycle packing of $G(S, P)$, the gray edges encode a one-to-one correspondence between genes from S and P. Two genes are correspondent if the vertices originated from them are connected by gray edges. Note that condition (ii) of cycle packing ensures that we do not have vertices originated from a gene of one genome connected with vertices from distinct genes of another genome.

In the Adjacency Graph Packing problem, given an adjacency graph $G(S, P)$, the goal is to find an alternating cycle packing $\mathcal{H}^*$ of G with maximum cardinality.

3   Random Packings

Our first approach to solve the Adjacency Graph Packing problem is the Random Packings (RP) heuristic. This heuristic randomly generates a set of **M** cycle packings and selects the one with maximum cardinality. Besides the strings S and P, the heuristic receives a parameter r indicating the number of cycles packings to be generated. The BFS_PACK procedure is responsible for the generation of each random cycle packing and is explained next.

A *partial alternating cycle packing* (or just *partial cycle packing*) $\mathcal{H}$ of the graph $G(S, P)$ is a set of disjoint alternating cycles, such that: (i) each vertex belongs to at most one alternating cycle; (ii) a gray edge $\{u, v\}$ belongs to a cycle iff it does not have a twin, its twin edge is also in some cycle, or it is possible to include a cycle in $\mathcal{H}$ containing its twin edge such that $\mathcal{H}$ remains a partial alternating cycle packing. In other words, a partial alternating cycle packing is an alternating cycle packing with more flexible restrictions allowing it to not cover all the vertices.

The BFS_PACK procedure initiates with a possibly empty partial cycle packing $\mathcal{H}$ and inserts cycles in it until enough cycles are inserted to cover all vertices, in this way, $\mathcal{H}$ becomes a cycle packing. Additionally, every insertion must ensure that $\mathcal{H}$ remains a partial cycle packing.

Each cycle is selected using a breadth-first search in the adjacency graph. At each iteration, the search for a new alternating cycle starts in a randomly chosen vertex v, such that v is not yet covered by any cycle of $\mathcal{H}$. During the search, the following restrictions must be followed to ensure that the new cycle does not violate the conditions of a partial cycle packing:

Algorithm 1: Random Packings

Data: balanced strings S and P, and number of cycle packings r
Result: cycle packing for $G(S, P)$

1   $\mathbf{M} \leftarrow \emptyset$
2   **while** $|\mathbf{M}| < r$ **do**
3      $\mathcal{H} \leftarrow \emptyset$
4   **while** $\mathcal{H}$ *does not cover all vertices of* $G(S, P)$ **do**
5         $v \leftarrow$ vertex of $G(S, P)$ not belonging to any cycle of $\mathbf{M}$
6         $C \leftarrow$ cycle resulting from a breadth-first search in $G(S, P)$, starting with v and following the necessary restrictions
7         $\mathcal{H} \leftarrow \mathcal{H} \cup C$
8      $\mathbf{M} \leftarrow \mathbf{M} \cup \{\mathcal{H}\}$

9   **return** *Cycle packing belonging to* $\mathbf{M}$ *with maximum cardinality*

- the search never uses an edge incident to a vertex belonging to a cycle already in $\mathcal{H}$ (to ensure that the first condition is met);
- If the current vertex has a twin edge of an edge belonging to a cycle in $\mathcal{H}$, that edge must be chosen (to ensure that the second condition is met).

The search stops when the first cycle is found. At that point the new cycle is added to $\mathcal{H}$ and, if $\mathcal{H}$ is not yet a cycle packing, a new search starts. Note that all vertices not covered by a cycle in $\mathcal{H}$ have an available gray and black edge, therefore, while the search does not find a cycle, it is possible to follow a black or gray edge to another vertex. Eventually, the search finds an alternating cycle of G not yet in $\mathcal{H}$.

Algorithm 1 presents a pseudocode of the Random Packings heuristic, where lines 4 to 7 show the BFS_PACK procedure. It is worth noting that the BFS_PACK procedure can be applied to any partial cycle packing (not necessarily empty) in order to obtain a cycle packing. Figure 3 shows an example of the BFS_PACK procedure.

4   Genetic Algorithm

In this section, we present our algorithm based on the Genetic Algorithm (GA) approach [10]. This strategy was inspired by the process of evolution, where, given an initial population, new generations arise from crossover and mutations of individuals from the previous generation. Furthermore, individuals more adapted to survive have a higher chance of participating in the crossovers. The idea behind genetic algorithms is to represent each solution as an individual and the value of that solution as the fitness of that individual. In the Adjacency Graph Packing problem, each individual is a cycle packing and their fitness is the size of the cycle packing.

Our genetic algorithm has three parameters: the total number of individuals to be generated r, the size of the population p, and the mutation rate m. The

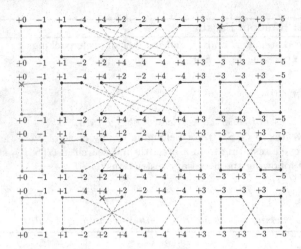

Fig. 3. Example of the BFS_PACK procedure applied in the adjacency graph of two balanced strings $S = (+1 + 4 - 2 - 4 - 3 + 3)$ and $P = (+1 + 2 - 4 + 4 - 3 - 3)$. The cycles obtained on each breadth-first search are marked with an × indicating the starting vertex of the search. Edges that can not be used in future searches are omitted from the figure.

algorithm starts with an initial population $\mathbf{P}_0$ of individuals generated by the Random Packings heuristic and, until we reach a total of r individuals, a new population $\mathbf{P}_{i+1}$ of individuals is produced using the individuals from the previous population $\mathbf{P}_i$. The algorithm takes the following steps in order to generate a new population:

1. Selection: At this step, the algorithm produces a set $\mathbf{S}$ with $p/2$ pairs of individuals to take part in the crossovers. Each individual of a pair is selected by tournament, in which two individuals are randomly chosen and the one with highest fitness is selected. A new pair is added to $\mathbf{S}$, where the two individuals are selected by two tournaments, until $\mathbf{S}$ has $p/2$ pairs. The random selection of individuals for the tournament allows repetitions, so an individual may belong to multiple pairs of $\mathbf{S}$.
2. Crossover: For each pair of $\mathbf{S}$, the algorithm applies twice the crossover operation to generate two new individuals. Given two cycle packings $\mathcal{H}$ and $\mathcal{H}'$ of an adjacency graph $G(S, P)$, a *crossover* of $\mathcal{H}$ and $\mathcal{H}'$ is a new cycle packing created by the following procedure. Let L and L' be two randomly ordered lists of the cycles from $\mathcal{H}$ and $\mathcal{H}'$, respectively. Starting with an empty set $\mathcal{H}''$, while $\mathcal{H}''$ is not a cycle packing and there are cycles in L or L', remove a cycle from L or L' (with a 50% probability to remove from each list, or 100% probability to remove from one list if the other is empty), and add it to $\mathcal{H}''$ if after the addition $\mathcal{H}''$ remains a partial cycle packing. If both lists L and L' are empty and $\mathcal{H}''$ is not yet a cycle packing, use the BFS_PACK procedure to complete the packing.

Algorithm 2: Genetic Algorithm

    Data: balanced strings S and P, number of cycle packings r, size of the
           population p, and mutation rate m
    Result: cycle packing for $G(S, P)$

1    $\mathbf{P}_0 \leftarrow p$ cycle packings generated by the Random Packings heuristic
2    $\mathbf{M} \leftarrow \mathbf{P}_0$
3    $i \leftarrow 0$
4   **while** $|\mathbf{M}| < r$ **do**
5        $\mathbf{S} \leftarrow$ Sequence of $p/2$ selected pairs of individuals from $\mathbf{P}_i$
6        $\mathbf{P}'_{i+1} \leftarrow$ Set of p new individuals created by crossover of individuals from $\mathbf{S}$
7        $\mathbf{P}''_{i+1} \leftarrow$ Set of p individuals obtained by mutation of individuals from $\mathbf{P}'_{i+1}$
8   **if** *the best individual of P_i is better than the worse individual of P''_{i+1}* **then**
9           $\mathbf{P}_{i+1} \leftarrow$ Set $\mathbf{P}''_{i+1}$ with its worse individual replaced with the best
             individual of P_i
10   **else**
11          $\mathbf{P}_{i+1} \leftarrow \mathbf{P}''_{i+1}$
12       $\mathbf{M} \leftarrow \mathbf{M} \cup \{\mathbf{P}_{i+1}\}$
13       $i \leftarrow i + 1$

14 **return** *Cycle packing of* $\mathbf{M}$ *with maximum cardinality*

---

3. Mutation: After the new individuals are generated, a mutation is applied to each individual in order to increase the diversity of the population. In the mutation of a cycle pack $\mathcal{H}''$, we remove each cycle of $\mathcal{H}''$ with probability m and create a new packing using the BFS_PACK procedure starting with the remaining cycles of $\mathcal{H}''$.
4. Elitism: As we are going to replace the old population with the new one, we cannot guarantee the quality of the new population. So, to ensure that at least the best individual is kept, we replace the individual with the lowest fitness from the new population with the individual with the highest fitness from the old population, if the old individual has a higher fitness than the new one.

Algorithm 2 presents a pseudocode for the Genetic Algorithm metaheuristic. Figure 4 exemplifies the generation of a new cycle packing after the crossover of two previous ones followed by a mutation.

5   Experimental Results

We create a dataset of simulated genomes to test the proposed heuristics. We also implemented and tested the SOAR framework developed by Chen *et al.* [3]. The SOAR framework comprises two steps: (i) the simplification of the input strings and (ii) the construction of a cycle packing using a heuristic similar to the BFS_PACK procedure. As a cycle packing for the simplified strings corresponds to a cycle packing for the original strings, we can compare it with the cycle packings returned by our heuristics.

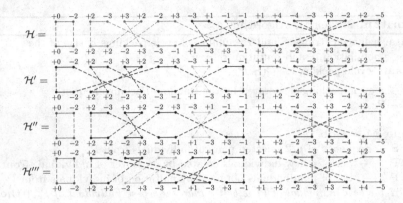

Fig. 4. The two cycle packings $\mathcal{H}$ and $\mathcal{H}'$ are combined by crossover resulting in the cycle packing $\mathcal{H}''$, where two cycles are copied from $\mathcal{H}$, two cycles are copied from $\mathcal{H}'$, and one cycle is inserted by the BFS_PACK procedure. Afterwards, the cycle packing $\mathcal{H}''$ suffers a mutation resulting in the cycle packing $\mathcal{H}'''$, where two cycles are removed and new cycles are inserted by the BFS_PACK procedure.

Our dataset has 1,000 pairs of strings, divided into 10 sets of 100 pairs. Each set uses an alphabet Σ of a specific size ($|\Sigma|$), such that sizes of alphabets range from 10 to 100 in intervals of 10. Each pair (S, T) of signed strings, representing the source (S) and target (T) strings, was generated as follows. We first construct the source string S of size 100 such that, at each position $i \in [1..100]$, we choose (with replacement and using a uniform distribution) a character from Σ. After that, the target string T is generated by shuffling S. Finally, each element from S and T randomly receives a sign ('+' or '-').

The Random Packing and Genetic Algorithm heuristics were implemented in C++, and the experiments were conducted on a PC equipped with a 2.3GHz Intel® Xeon® CPU E5-2650 v3, with 40 cores and 32 GB of RAM, running Ubuntu 18.04.2.

Table 1 shows the average cycle packing size returned and the average execution time (in seconds) of the Genetic Algorithm heuristic with $r = 1000$, $p = 100$, and $m = 0.5$ (GA-1k), the Genetic Algorithm heuristic with $r = 10000$, $p = 100$, and $m = 0.5$ (GA-10k), the Random Packing heuristic with $r = 1000$ (RP-1k), and the SOAR framework (SOAR).

We can see in Table 1 that with $r = 1000$ cycle packings both RP and GA heuristics were able to find cycle packings larger than the ones encountered by the SOAR framework. Also, the cycle packings returned by the GA heuristic were, on average, larger than the ones returned by the RP heuristic. Regarding the GA heuristic with $r = 10000$, the average size of the best cycle packing returned was at least 30% higher than the average size of the cycle packing returned by the SOAR framework.

Table 1. Average cycle packing size returned (CPS) and average execution time (in seconds) of our heuristics and the SOAR framework.

$|\Sigma|$	RP-1k		GA-1k		GA-10k		SOAR	
	CPS	Time	CPS	Time	CPS	Time	CPS	Time
10	35.80	73.15	35.91	32.09	43.71	245.20	32.74	0.03
20	28.47	55.13	28.99	24.48	35.03	183.16	25.81	0.02
30	24.50	47.05	25.34	21.63	29.32	156.66	20.99	0.02
40	21.84	40.86	22.87	19.49	25.43	137.59	18.18	0.02
50	19.97	38.20	20.91	18.01	22.52	134.03	15.88	0.02
60	18.54	35.93	19.46	17.12	20.70	126.23	1.4.49	0.02
70	17.49	33.10	18.36	16.05	19.08	122.31	13.26	0.02
80	16.64	32.44	17.44	15.37	18.09	121.12	12.57	0.02
90	15.79	31.15	16.30	14.51	16.80	116.35	11.45	0.02
100	15.25	30.29	15.81	14.24	16.17	113.59	11.17	0.02

Concerning the execution times from Table 1, we can see that obtaining better packings (i.e., the ones generated by our heuristics instead of SOAR framework) comes at a cost. Besides, the GA heuristic with $r = 1000$ is less time-consuming than the RP heuristic using the same value of r and it returns cycle packings that are on average better than RP. If time is limited, the GA heuristic with $r = 1000$ has a good trade-off between running time and the quality of the solution. If time is not a problem, the GA heuristic with $r = 10000$ returns cycle packings that have on average 11% more cycles than the ones returned by GA using $r = 1000$.

5.1 Applications with the Reversal Distance

A *reversal* is a large-scale mutation that inverts a subsequence of genes (and, consequently, their orientations). Formally, a reversal $\rho(i,j)$, with $1 \leq i \leq j \leq n$, when applied to a given signed string $S = (S_1 \ldots S_{i-1} S_i \ldots S_j S_{j+1} \ldots S_n)$ generates a new string $S' = S \cdot \rho(i,j) = (S_1 \ldots S_{i-1} \overline{-S_j \ldots -S_i} S_{j+1} \ldots S_n)$.

The *Sorting by Reversals* problem (SbR) seeks the minimum number of reversals required to transform one genome into the other, which is defined as the *reversal distance*.

As explained in Sect. 1, the reversal distance can be obtained in polynomial time if orientations are known and there are no duplicated genes [7]. This is due to the fact that there is a unique cycle packing in its adjacency graph. In the case of genomes with repeated genes, we can use any cycle packing obtained as an upper bound for the reversal distance, since each packing defines a one-to-one relation between genes of same label.

Table 2 shows the average reversal distance using the best cycle packing returned by the Genetic Algorithm heuristic with $r = 1000$ (column GA-1k)

Table 2. Average reversal distance when using the cycle packing provided by our heuristics and by the SOAR framework.

$|\Sigma|$	RP–1k	GA–1k	GA–10k	SOAR
10	65.20	65.09	57.29	68.13
20	72.53	72.01	65.97	75.12
30	76.50	75.66	71.68	79.94
40	79.16	78.13	75.57	82.79
50	81.03	80.09	78.48	85.10
60	82.46	81.54	80.30	86.47
70	83.51	82.64	81.92	87.70
80	84.36	83.56	82.91	88.41
90	85.21	84.70	84.20	89.53
100	85.75	85.19	84.83	89.82

and with $r = 10000$ (column GA-10k), using the best cycle packing returned by the Random Packing heuristic with $r = 1000$ (column RP-1k), and using the cycle packing obtained from the SOAR framework (column SOAR). We can see the distances obtained with the SOAR algorithm are on average greater than those obtained by the proposed heuristics, regardless of the size of the alphabet. For instance, the reversal distances using the cycle packing from the RP heuristic were on average 4.4% lower than using the cycle packing provided by SOAR framework, and the reversal distance using the cycle packing from the GA heuristic with $r = 1000$ and $r = 10000$ were on average 5.3% and 8.7% lower than the SOAR ones. Besides, the reversal distances using cycle packings from the GA heuristic with both values of r were also consistently lower, on average, than the reversal distances using cycle packings from the RP heuristic.

5.2   Experiments with Real Biological Data

To validate our algorithm with real biological data, we applied our Genetic Algorithm heuristic for genomes from the Cyanorak 2.1 [6] system. Cyanorak 2.1 has a total of 97 genomes, being 51 Synechococcus, 43 Prochlorococcus, and 3 Cyanobium. The genomes sizes (number of genes) range from 1834 to 4391. Considering the ratio between the number of replicated genes and the total number of genes in each genome, we have 0.35%, 1.93%, and 10.13% of minimum, average, and maximum ratios, respectively. Considering the gene with the highest occurrence in each genome, we have the values of 2, 7.8, and 92 for minimum, average, and maximum, respectively. From these values, we see that the number of replicated genes compared to the size of the genomes is relatively small.

As our Genetic Algorithm heuristic can only be applied in balanced strings, we performed a pre-processing step on each genome pair removing all non-common genes and extra gene copies from each genome. For each gene present in

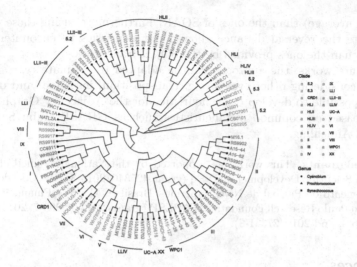

Fig. 5. Phylogenetic tree resulting from the reversal distances obtained with the Genetic Algorithm heuristic.

both genomes, we kept the first s copies from each genome, where s is the minimum occurrence of the gene in the genome pair, and we removed the remaining copies. Next, we applied the Genetic Algorithm heuristic (with $r = 100$, $p = 10$, and $m = 0.5$) to estimate the reversal distances for each pair of genomes.

The tests took, on average, 19 s for each pair of genomes. Using the resulting distances, we constructed a phylogenetic tree with the Circular Order Reconstruction method [9], which is shown in Fig. 5[1]. We can note that the genomes of organisms are, in general, grouped according to their clade. Our phylogenetic tree is considerably similar to the one presented by Laurence *et al.* [6], as pointed by the P-value obtained using the approach proposed by de Vienne, Giraud, and Martin [5] based on the maximum agreement subtrees (MAST). This P-value corresponds to the probability of the trees being unrelated. We obtained a MAST with 42 leaves and P-value = 3.47×10^{-19}.

6 Conclusion

In this work, we investigated the Adjacency Graph Packing problem, which is a NP-hard problem when genomes have repeated genes. The goal of this problem is to find a cycle packing that has as many cycles as possible. We developed two heuristics, one that uses a random approach and one based on the Genetic Algorithm approach.

We implemented these heuristics and tested them in simulated and real datasets. We also compared both heuristics with the SOAR framework. Although more time-consuming, our heuristics were able to find cycle packings with more

---

[1] Illustration created using `treeio` R package [14].

cycles (on average) than the ones of SOAR. Furthermore, using these packings to estimate the reversal distance allowed our heuristics to encounter shorter distances than the ones provided by the SOAR framework.

As future works, the approach can be adapted to estimate rearrangement distances considering other events (e.g. transposition, insertion, and deletion). It is also possible to develop other heuristics for the Adjacency Graph Packing problem based on common metaheuristics such as GRASP, Tabu Seach, and Simulated Annealing.

Acknowledgments. This work was supported by the National Council of Technological and Scientific Development, CNPq (grant 425340/2016-3), the Coordenação de Aperfeiçoamento de Pessoal de Nível Superior - Brasil (CAPES) - Finance Code 001, and the São Paulo Research Foundation, FAPESP (grants 2013/08293-7, 2015/11937-9, 2017/12646-3, and 2019/27331-3).

References

1. Bafna, V., Pevzner, P.A.: Genome rearrangements and sorting by reversals. SIAM J. Comput. **25**(2), 272–289 (1996)
2. Bergeron, A., Mixtacki, J., Stoye, J.: A unifying view of genome rearrangements. In: Bücher, P., Moret, B.M.E. (eds.) WABI 2006. LNCS, vol. 4175, pp. 163–173. Springer, Heidelberg (2006). https://doi.org/10.1007/11851561_16
3. Chen, X., et al.: Assignment of orthologous genes via genome rearrangement. IEEE/ACM Trans. Comput. Biol. Bioinf. **2**(4), 302–315 (2005)
4. Christie, D.A.: Genome Rearrangement Problems. Ph.D. thesis, Department of Computing Science, University of Glasgow (1998)
5. De Vienne, D.M., Giraud, T., Martin, O.C.: A congruence index for testing topological similarity between trees. Bioinformatics **23**(23), 3119–3124 (2007)
6. Garczarek, L., et al.: Cyanorak v2. 1: a scalable information system dedicated to the visualization and expert curation of marine and brackish picocyanobacteria genomes. Nucleic Acids Res. **1** (2020)
7. Hannenhalli, S., Pevzner, P.A.: Transforming cabbage into turnip: polynomial algorithm for sorting signed permutations by reversals. J. ACM **46**(1), 1–27 (1999)
8. Kahn, C., Raphael, B.: Analysis of segmental duplications via duplication distance. Bioinformatics **24**(16), i133–i138 (2008)
9. Makarenkov, V., Leclerc, B.: Circular orders of tree metrics, and their uses for the reconstruction and fitting of phylogenetic trees. In: DIMACS Series in Discrete Mathematics and Theoretical Computer Science, pp. 183–208. American Mathematical Society (1997)
10. Mitchell, M.: Introduction to Genetic Algorithms. Springer, Berlin Heidelberg, Cambridge, MA, USA (2008)
11. Pinheiro, P.O., Alexandrino, A.O., Oliveira, A.R., de Souza, C.C., Dias, Z.: Heuristics for breakpoint graph decomposition with applications in genome rearrangement problems. In: BSB 2020. LNCS, vol. 12558, pp. 129–140. Springer, Cham (2020). https://doi.org/10.1007/978-3-030-65775-8_12
12. Radcliffe, A.J., Scott, A.D., Wilmer, E.L.: Reversals and transpositions over finite alphabets. SIAM J. Discrete Math. **19**(1), 224–244 (2005)

13. Shao, M., Lin, Y., Moret, B.M.: An exact algorithm to compute the double-cut-and-join distance for genomes with duplicate genes. J. Comput. Biol. **22**(5), 425–435 (2015)
14. Wang, L.G., et al.: Treeio: an R package for phylogenetic tree input and output with richly annotated and associated data. Mol. Biol. Evol. **37**(2), 599–603 (2020)
15. Willing, E., Stoye, J., Braga, M.D.: Computing the inversion-indel distance. IEEE/ACM Trans. Comput. Biol. Bioinf. (2020)

PIMBA: A PIpeline for MetaBarcoding Analysis

Renato R. M. Oliveira[1,2]([⊠]), Raíssa Silva[1], Gisele L. Nunes[1],
and Guilherme Oliveira[1]([⊠])

[1] Instituto Tecnológico Vale, Belém, Pará, Brazil
guilherme.oliveira@itv.org
[2] Programa de Pós-Graduação em Bioinformática, Instituto de Ciências Biológicas,
Universidade Federal de Minas Gerais, Belo Horizonte, Minas Gerais, Brazil

Abstract. DNA metabarcoding is an emerging monitoring method capable of
assessing biodiversity from environmental samples (eDNA). Advances in computational tools have been required due to the increase of Next-Generation Sequencing data. Tools for DNA metabarcoding analysis, such as MOTHUR, QIIME,
Obitools, PEMA, and mBRAVE have been widely used in ecological studies,
however, some difficulties are encountered when there is a need to use custom databases. Here we present PIMBA, a PIpeline for MetaBarcoding Analysis, which allows the use of customized databases, as well as other reference
databases used by the software mentioned here. PIMBA is an open-source and
user-friendly pipeline that consolidates all analyses in just three command lines.
PIMBA's implementation is available at https://github.com/reinator/pimba.

Keywords: DNA metabarcoding · Flexible pipeline · OTU · ASV

1 Introduction

DNA metabarcoding is a powerful tool that has been widely used for biodiversity monitoring and ecosystem assessment from environmental DNA samples (eDNA). The technique based on high-throughput sequencing (HTS) allows the multispecies detection
from specific molecular markers in a group (plants, animals, fungi, and bacteria) [1].
Next-Generation Sequencing (NGS) results in millions of DNA sequences (reads) that
allow deciphering the genetic code of species, answering taxonomic and functional questions. Nowadays, different sequencing platforms are available and can generate either
paired-end or single-end reads. The technologies associated with paired-end reads allow
to sequence a pool of different samples and automatically demultiplex them by using
different indexes. Such a pool of samples can also be achieved with single-end technologies, but they still did not automatize the demultiplexing steps. If single-end reads
are being sequenced for a metabarcoding analysis, the pool of samples might be done
by using either single or dual-indexes multiplexing, this latter allowing to pool a larger
number of samples in the same sequencing run, reducing costs and time.

The recent increase of NGS data has caused the development of new tools for DNA
metabarcoding analysis, making the metabarcoding method more accessible and user-
friendly [2]. Mothur [3], Qiime [4], Obitools [5], mBRAVE [6], and PEMA [7] are

P. F. Stadler et al. (Eds.): BSB 2021, LNBI 13063, pp. 106–116, 2021.
https://doi.org/10.1007/978-3-030-91814-9_10

currently the most used tools for metabarcoding analysis. Most pipelines use operational taxonomic units (OTUs) as the clustering method, except Pema which works for both OTU clustering (1) and Amplicon Sequence Variants (ASVs) inference (2). In the first approach (1) the reads are grouped into OTUs to differentiate species or taxa based on the similarity of sequences [8, 9]. A similarity of 97% is commonly used as a cutoff, but this value depends on the group to be evaluated [10]. The second approach (2) infers ASVs, which are reads that differ in 1 nucleotide (or more than) [11].

The biggest restriction of these pipelines is to make it difficult of using a customized database for taxonomy assignments. Among the available tools, Mothur is useful when analyzing 16S/18S rRNA, Influenza viral, and fungal ITS regions, using Greengenes [12], Influenza Virus, and SILVA [13] databases, respectively. Qiime (and even its updated version, Qiime2) is optimized to analyze metabarcoding data from 16S rRNA, 18S rRNA, and fungal ITS marker genes, using Greengenes, SILVA, and UNITE [14] databases, respectively. Qiime2 allows the user to train a classifier with a NaiveBayes model, but they report tests by using only a 16S example and with some constraints to the use of this classifier with other marker genes. Obitools is optimized to analyze data from 16S (SILVA and PR2) and it also allows the use of the NCBI database for taxonomic assignment. mBRAVE is optimized to use only the BOLD [15] database as a reference, allowing the researcher to use a personalized database only after BOLD submission. PEMA allows the analysis of metabarcoding data from 16S/18S rRNA, fungal ITS, and metazoan COI, using SILVA, UNITE, and MIDORI [16] databases, respectively.

To allow the researcher to use a customized database in the metabarcoding analyses as well as reference database such as NCBI/Genbank, we developed PIMBA, a PIpeline for MetaBarcoding Analysis, which adapts the Qiime/BMP [17] pipeline for OTUs clustering with additional and optional OTU corrections based on the algorithm LULU [18]. PIMBA accepts both single and paired-end reads, with both single and dual-index. PIMBA also allows inferring ASVs using Swarm [19]. A preliminary abundance and diversity analysis are also automatically delivered. The main innovation of this pipeline is, in just three command lines, the ease of using both standard and customized databases, minimizing errors in taxonomic assignments.

2 Implementation

PIMBA is fully containerized in docker images, being more platform-independent and easy to maintain and update. Besides implementing all the features provided by the other metabarcoding tools, PIMBA also allows the user to apply different databases and not only those commonly used by most of the available software. PIMBA can be used with single or paired-end reads and is divided into three steps: (1) preprocessing, which promotes the demultiplexing and quality treatment of reads (Fig. 1A); (2) taxonomy assignment, in which reads are clustered into OTUs or ASVs are inferred, along with errors correction (Fig. 1B), and (3) plotting, in which alfa and beta diversity plots are built by Phyloseq [20], including rarefaction curves and Principal Coordinates Analysis (PCoA), respectively. A metadata file is required for this last step (all PIMBA commands are available at https://github.com/reinator/pimba).

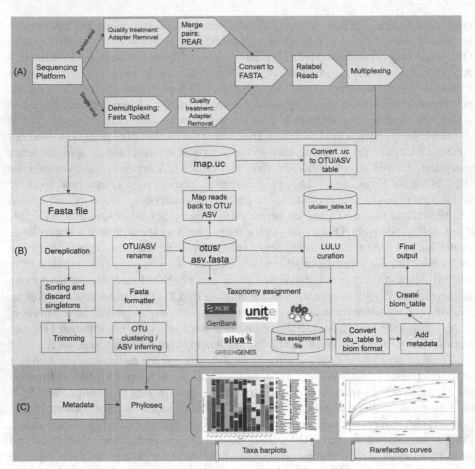

Fig. 1. PIMBA workflow. (A) preprocessing, (B) OTU clustering or ASV inference, and (C) plotting.

2.1 Preprocessing

PIMBA (pimba_prepare) can process either single-end or paired-end reads. Depending on the sequencing strategy, a few steps for demultiplexing or merging reads are needed. For paired-end reads, first AdapterRemoval v2.2.3 [21] will trim the adapters used in the sequencing (-mm 5, allowing 10% difference in the adapters sequence) with additional quality treatment, by using a 10bp window (--trimwindows), removing all reads with mean quality below PHRED 20 (--minquality) and with length less than 100bp (--minlength). Then, all read pairs will be merged with PEAR [22], using default parameters.

For single-end reads, a demultiplex step is performed. PIMBA allows the demultiplex of both single and dual-index libraries. In both cases, PIMBA will use Fastx-Toolkit (http://hannonlab.cshl.edu/fastx_toolkit/) to detect the 5' or 3' index (in the case of single-index) or the 5' and 3' indexes (in the case of dual-index), generating at the end all the demultiplexed reads. Then, AdapterRemoval is used to perform the quality treatment with the same parameters as mentioned for paired-end reads.

Both high-quality paired-end and single-end demultiplexed reads are converted to FASTA with Prinseq v0.20.4 [23], relabeled and concatenated to a single FASTA file, that will be used in the next step, for OTU clustering or ASV inference. All the steps above for preprocessing (Fig. 1A) illumina paired-end datasets with 10 threads, for example, can be run in the following command-line:./pimba_prepare.sh illumina < rawdata_dir > < output_reads > < num_threads > < adapters.txt > < min_length > < min_phred > (e.g.: pimba_prepare.sh illumina rawdata_dir/ AllSamples_16S_hqdata 10 adapters.txt 100 20).

2.2 Taxonomy Assignment

With the multiplexed FASTA file resulting from the preprocessing step, PIMBA (pimba_run) will use VSEARCH v2.15.2 [24] to dereplicate, discard singletons and trim the reads to a given length (−1 0, if no trim is desired). Then, depending on the approach, PIMBA will cluster the reads into OTUs (-w out) at a given similarity threshold (-s) using VSEARCH or infer the ASVs with Swarm (-w asv), accepting difference in only one nucleotide. For both OTUs/ASVs, PIMBA will use VSEARCH to remove chimeras (--uchime_denovo), Fastx_toolkit to format the FASTA file, and a Perl script from BMP to rename the OTUs/ASVs. VSEARCH will also be used to map back the reads to the OTUs/ASVs and then a script from QIIME will be used to generate an OTU/ASV table. PIMBA will optionally use LULU to curate all the found OTUs/ASVs (-x).

Depending on the marker gene the user is analyzing (-g), PIMBA will use different databases to taxonomically assign the OTUs/ASVs. To work properly, the user will need to pass a database file as a parameter (-d), where the location from the desired database will be set. Currently, PIMBA allows the analysis of 16S rRNA (SILVA, Greengenes, RDP [25] or Genbank [26]), fungal ITS (UNITE or Genbank), and for any other desired marker gene (e.g., metazoan COI, plant ITS) with BLAST [27] assignment to the Genbank database. Also, PIMBA allows the user to generate a customized database for assignments. (see https://github.com/reinator/pimba). When the user desires to use the SILVA, Greengenes, RDP, or UNITE databases, PIMBA will use scripts adapted from QIIME/BMP pipeline, and the user will also need to define the similarity for the assignment (-a). In the case of analyzing fungal ITS, PIMBA will also use ITSx [28] to discard the ribosomal regions flanking the ITS regions. When the user desires to use the NCBI Genbank database, a set of PIMBA scripts will be used, and besides the assignment similarity, the user will need to define the minimum alignment coverage (-c), the maximum e-value allowed (-e), and the number of hits per sequence that BLAST needs to return (-h). When -h is 1, only the best hit is returned. If -h is greater than 1, PIMBA will perform a voting system to properly assign the taxonomy. In case of a tie, the taxon with greater similarity will be chosen.

Finally, PIMBA will use Biom v2.1.10 [29] to convert the OTU/ASV table to biom format and add the taxonomy assignments, generating a summarized biom table file, needed in the next step.

All the steps above for running taxonomy assignment (Fig. 1B) can be run in the following command-line:./pimba_run.sh -i < input_reads > -o < output_dir > -w < approach > -s < otu_similarity > -a < assign_similarity > -c < coverage > -l < otu_length > -h < hits_per_subject > -g < marker_gene > -t < num_threads > -e < E-value > -d < databases.txt > -x < lulu > (e.g.: pimba_run.sh -i 16S_hqdata.fasta -o run_OTU_NCBI -w otu -s 0.97 -a 0.9 -c 0.9 -l 200 -h 5 -g 16S-NCBI -t 10 -d databases.txt).

2.3 Plotting

In the end, PIMBA (pimba_plot) will use Phyloseq to plot alpha and beta diversity results, such as rarefaction curves and PCoA plots. The user only needs to give as parameters the OTU/ASV table (-t), the taxonomy assignment file (-a), and a metadata file (-m). Depending on the metadata file, the user will also be able to group the results according to a given attribute from the samples (-g). To perform the plotting (Fig. 1C), the following command-line can be used: pimba_plot.sh -t < otu_table > -a < tax_assignment > -m < metadata > -g < group_by > (e.g.: pimba_plot.sh -t 16S_otu_table.txt -a 16S_otus_tax_assignments.txt -m mapping_file.csv -g Description).

3 Results and Discussion

To demonstrate that PIMBA is effective at analyzing a metabarcoding dataset, we used the same benchmark used by PEMA [7]: three mock communities sequencing. PIMBA's results are also being compared to the results presented in the PEMA publication.

From the mock community sequencing, the first dataset is from the 16S rRNA gene, comprising 20 bacterial species [30]. The second is a dataset from fungal ITS, comprising 19 fungal species [31]. The third is a dataset from metazoan COI amplicons, comprising 14 species [32]. Information regarding the datasets mentioned above is summarized in Table 1.

To evaluate the mock communities' results, we ran an extensive comparative benchmark, varying parameters such as truncation length, minimum assignment similarity, taxon database, and strategy (with OTU clustering or ASV inference). For each test, we calculated the True-Positives (TP), False-Positives (FP), and False-Negatives (FN) obtained by PIMBA, at both Genus and Species levels. Then, we were able to calculate precision (to check how many correct results PIMBA returned), recall (to check how much of the known taxa in the mock communities PIMBA can recover), and F1 score (which combines the precision and recall values) [33]. All TP, FP, FN, and F1 values obtained in the tests we performed are available at https://github.com/reinator/pimba.

For all tests, we fixed the cluster similarity for OTUs (-s 0.97) and maximum difference for ASV (-d 1). Besides, only taxa existing in all replicates were considered as a hit, except for the COI dataset, where we accepted as a hit, taxon occurring in at least two replicates, given its low depth. We also decided not to run LULU curation, as we saw that in all tests, the F1 scores were lower than when LULU was not used. In the next sections, the results from the mock datasets will be described and discussed.

Table 1. Mock community datasets and accessions. Total reads, bases, and sequencing read length are also shown.

Marker gene	SRA	Total reads	Total bases (Mb)	Read length (bp)
16S	SRR3163904 SRR3163905 SRR3163906	895,113	471.3	2 × 300
Fungal ITS	SRR5838515 SRR5838516 SRR5838522	162,841	81.5	2 × 250
Metazoan COI	ERR2181459 ERR2181468 ERR2181466	228,019	113.8	2 × 300

3.1 16S rRNA Mock Community

The 16S rRNA mock community was sequenced with Illumina MiSeq using the v3 reagent kit (2x300 cycles), targeting the V4 region (~252 bp) [30]. After quality treatment and pair merging, a total of 810,981 amplicon reads were used as input to pimba_run. We varied the strategy (OTU or ASV), the minimum assignment similarity (0.90, 0.97, and 0.99), the truncation length (200 bp, 250 bp), and the taxon database (SILVA or Genbank/NCBI). The F1 scores obtained at the Genus level are shown in Table 2. The best F1 scores are highlighted in all the tables that follow.

Table 2. F1 scores for each one of PIMBA's 16S rRNA results at Genus level, when varying assignment similarity, truncation length, strategy, and taxon database.

Min assign similarity	0.90		0.97		0.99	
Truncation length	200 bp	250 bp	200 bp	250 bp	200 bp	250 bp
OTU - SILVA	0.89	0.89	0.89	0.91	0.88	0.90
ASV - SILVA	0.95	0.95	0.95	0.93	**0.98**	0.95
OTU – Genbank	0.90	0.93	0.90	0.93	0.90	0.93
ASV - Genbank	0.93	0.95	0.88	0.95	0.90	0.95

PIMBA performed better (F1 score = 0.98) when running ASV inference, truncating the sequences at 200bp and assigning similarity only above 99% against the SILVA database. This configuration returned only one false positive (*Prevotella*) and recovered all 20 bacterial taxa, being 1.18-fold better than PEMA's results (F1 = 0.83, see [7]) when analyzing the same dataset. At the Species level, PIMBA performed better when running OTU clustering at 250bp and assigning the taxa at any of the selected similarities (Table 3).

Table 3. F1 scores for each one of PIMBA's 16S rRNA results at Species-level, when varying assignment similarity, truncation length, strategy, and taxon database.

Min assign similarity	0.90		0.97		0.99	
Truncation length	200 bp	250 bp	200 bp	250 bp	200 bp	250 bp
OTU - SILVA	0.28	0.39	0.21	0.33	0.22	0.37
ASV - SILVA	0.58	0.58	0.41	0.61	0.60	0.59
OTU – Genbank	0.78	**0.80**	0.78	**0.80**	0.78	**0.80**
ASV - Genbank	0.62	0.68	0.61	0.68	0.60	0.68

PIMBA recovered 17 species of the 20 bacterial taxa in the mock community when used with OTU strategy, being 5.6-fold better than PEMA, which recovered only 3 species.

3.2 Fungal ITS Mock Community

The fungal mock community targeted the ITS2 region (~327bp, ± 40.1) [34] and was sequenced with Illumina MiSeq, using v2 reagent kit (2 × 250 cycles) [31]. The pimba_prepare script outputted a total of 155,691 amplicon reads, which were used by pimba_run. We compared the results by varying the strategy (OTU or ASV), the minimum assign similarity (0.90, 0.95, and 0.97), the truncation length (100 bp, 130 bp, and 160 bp), and the taxon database (UNITE or Genbank/NCBI). The F1 scores obtained at the Genus levels are shown in Table 4.

Table 4. F1 scores for each one of PIMBA's ITS results at Genus level, when varying assignment similarity, truncation length, strategy, and taxon database.

Min assign similarity	0.90			0.95			0.97		
Truncation length	100 bp	130 bp	160 bp	100 bp	130 bp	160 bp	100 bp	130 bp	160 bp
OTU - UNITE	0.85	0.88	0.64	0.88	0.85	0.64	0.85	0.88	0.64
ASV -UNITE	0.85	0.85	0.59	0.85	0.85	0.64	0.81	0.81	0.69
OTU - Genbank	**0.94**	**0.94**	0.85	**0.94**	**0.94**	0.85	**0.94**	**0.94**	**0.94**
ASV - Genbank	**0.94**	**0.94**	0.85	**0.94**	**0.94**	0.85	**0.94**	0.94	0.85

For ITS, PIMBA performed better (F1 = 0.94) when using the Genbank database for taxonomy assignment, being 1.09-fold better than PEMA (F1 = 0.86, see [7]).

Both OTU and ASV strategies used by PIMBA had the same F1 scores in almost all configurations, except for truncation at 160bp, with 0.97 of assignment similarity, where the OTU strategy outperformed ASV's. At the Species level, PIMBA performed better when running OTU clustering at 100bp and assigning the taxa at any of the selected similarities (Table 5) using the Genbank database.

Table 5. F1 scores for each one of PIMBA's ITS results at Species-level, when varying assignment similarity, truncation length, strategy, and taxon database.

Min assign similarity	0.90			0.95			0.97		
Truncation length	100 bp	130 bp	160 bp	100 bp	130 bp	160 bp	100 bp	130 bp	160 bp
OTU - UNITE	038	0.37	0.33	0.44	0.38	0.32	0.43	0.36	0.31
ASV -UNITE	0.38	0.37	0.26	0.43	0.38	0.33	0.37	0.36	0.31
OTU - Genbank	**0.74**	0.72	0.63	**0.74**	0.72	0.61	**0.74**	0.72	0.72
ASV - Genbank	0.67	0.67	0.61	065	0.67	061	0.65	0.67	0.63

PIMBA recovered 14 species of the 19 bacterial taxa in the mock community when using either OTU or ASV strategy and truncating at 100 bp, being 2.8-fold better than PEMA, which recovered only 5 species. However, the number of false positives increased when the ASV strategy was used (10 False Positives), in comparison to OTU (5 False Positives).

3.3 Metazoan COI Mock Community

This dataset comprises a 3' region from the Cytochrome oxidase I gene (~450bp), sequenced with Illumina MiSeq, using v2 reagent kit (2x250 cycles)[32]. After performing preprocessing in the paired-end raw data, pimba_prepare outputted a total of 141,283 amplicon reads. We compared the results by varying the strategy (OTU or ASV), the minimum assign similarity (0.97, 0.98, and 0.99), the truncation length (250 bp, 350 bp, and 450 bp). PIMBA does not use a specific database for metazoan COI, so the taxon database used was Genbank/NCBI. The F1 scores obtained at the Genus levels and species levels were the same and are shown in Table 6.

PIMBA's performance was quite homogenous when varying OTU and ASV, getting an incredible F1 score of 1 in almost all configurations, being 1.35-fold better than PEMA (F1 = 0.74, see [7]). PIMBA recovered all 13 invertebrate species from the mock community.

Table 6. F1 scores for each one of PIMBA's COI results at Genus and Species-level, when varying assignment similarity, truncation length, strategy, and taxon database.

Min assign similarity	0.97			0.98			0.99		
Truncation length	250 bp	350 bp	450 bp	250 bp	350 bp	450 bp	250 bp	350 bp	450 bp
OTU	0.96	1	1	0.96	1	1	0.96	1	1
ASV	1	1	0.96	1	1	1	1	0.96	1

4 Conclusion

In contrast to the pipelines mentioned above, PIMBA allows the use of some specific or commonly used databases, such as Genbank, for taxonomy assignment. This feature is of paramount importance when there is a need to work with private and non-public databases. Another advantage of PIMBA is the freedom to use different forms of grouping sequences (ASVs or OTUs) within the same pipeline (most available pipelines apply a unique grouping approach). Regarding the results, it was possible to see how accurate PIMBA is in obtaining taxon for both Genus and Species levels and how flexible it is in the use of different strategies, parameters, and databases. Using as a comparison the PEMA pipeline, which applies similar strategies to PIMBA, we show that our results (both OTUs and ASVs) were superior concerning the expected taxonomy since we used a mock community as a dataset. Regarding the choice of the best grouping strategy (OTU or AVS) for our dataset, OTU presented a better resolution at the Species level, especially when using the Genbank database, while the ASV approach showed better results for analysis at the level of Genus.

References

1. Creer, S., et al.: The ecologist's field guide to sequence-based identification of biodiversity. Meth. Ecol. Evol. **7**, 1008–1018 (2016). https://doi.org/10.1111/2041-210X.12574
2. Alberdi, A., Aizpurua, O., Gilbert, M.T.P., Bohmann, K.: Scrutinizing key steps for reliable metabarcoding of environmental samples. Meth. Ecol. Evol. **9**, 134–147 (2018). https://doi.org/10.1111/2041-210X.12849
3. Schloss, P.D., et al.: Introducing mothur: Open-source, platform-independent, community-supported software for describing and comparing microbial communities. Appl. Environ. Microbiol. **75**, 7537–7541 (2009). https://doi.org/10.1128/AEM.01541-09
4. Caporaso, J.G., et al.: QIIME allows analysis of high-throughput community sequencing data. Nat. Meth. **7**, 335–336 (2010). https://doi.org/10.1038/nmeth.f.303
5. Boyer, F., Mercier, C., Bonin, A., Le Bras, Y., Taberlet, P., Coissac, E.: Obitools : a unix -inspired software package for DNA metabarcoding. Mol. Ecol. Resour. **16**, 176–182 (2016). https://doi.org/10.1111/1755-0998.12428
6. Ratnasingham, S.: mBRAVE: the multiplex barcode research and visualization environment. Biodivers. Inf. Sci. Stand. **3**, e37986 (2019). https://doi.org/10.3897/biss.3.37986

7. Zafeiropoulos, H., et al.: PEMA: a flexible pipeline for environmental DNA metabarcoding analysis of the 16S/18S ribosomal RNA, ITS, and COI marker genes. Gigascience **9**, 1–12 (2020). https://doi.org/10.1093/GIGASCIENCE/GIAA022

8. Cristescu, M.E.: From barcoding single individuals to metabarcoding biological communities: towards an integrative approach to the study of global biodiversity. Trends Ecol. Evol. **29**(10), 566-571 (2014). https://doi.org/10.1016/j.tree.2014.08.001

9. Hering, D., et al.: Implementation options for DNA-based identification into ecological status assessment under the European water framework directive. Water Res. **138**, 192–205 (2018). https://doi.org/10.1016/j.watres.2018.03.003

10. Deiner, K., et al.: Environmental DNA metabarcoding: transforming how we survey animal and plant communities. Mol. Ecol. **26**, 5872–5895 (2017). https://doi.org/10.1111/mec.14350

11. Callahan, B.J., McMurdie, P.J., Holmes, S.P.: Exact sequence variants should replace operational taxonomic units in marker-gene data analysis. ISME J. **11**(12), 2639–2643 (2017). https://doi.org/10.1038/ismej.2017.119

12. DeSantis, T.Z., et al.: Greengenes, a chimera-checked 16S rRNA gene database and workbench compatible with ARB. Appl. Environ. Microbiol. **72**, 5069–5072 (2006). https://doi.org/10.1128/AEM.03006-05

13. Quast, C., et al.: The SILVA ribosomal RNA gene database project: improved data processing and web-based tools. Nucleic Acids Res. **41**, D590–D596 (2013). https://doi.org/10.1093/nar/gks1219

14. Abarenkov, K., et al.: The UNITE database for molecular identification of fungi – recent updates and future perspectives. https://www.jstor.org/stable/27797548. (2010). https://doi.org/10.2307/27797548

15. Ratnasingham, S., Hebert, P.D.N.: BARCODING: bold: the barcode of life data system (http://www.barcodinglife.org). Mol. Ecol. Notes. **7**, 355–364 (2007). https://doi.org/10.1111/j.1471-8286.2007.01678.x

16. Machida, R.J., Leray, M., Ho, S.-L., Knowlton, N.: Metazoan mitochondrial gene sequence reference datasets for taxonomic assignment of environmental samples. Sci. Data **41**(4), 1–7 (2017). https://doi.org/10.1038/sdata.2017.27

17. Pylro, V.S., et al.: Brazilian microbiome project: revealing the unexplored microbial diversity—challenges and prospects. Microb. Ecol. **67**(2), 237–241 (2013). https://doi.org/10.1007/s00248-013-0302-4

18. Frøslev, T.G., et al.: Algorithm for post-clustering curation of DNA amplicon data yields reliable biodiversity estimates. Nat. Commun. **8**, 1–11 (2017). https://doi.org/10.1038/s41467-017-01312-x

19. Mahé, F., Rognes, T., Quince, C., de Vargas, C., Dunthorn, M.: Swarm v2: highly-scalable and high-resolution amplicon clustering. Peer J. **3**, e1420 (2015). https://doi.org/10.7717/PEERJ.1420

20. McMurdie, P.J., Holmes, S.: phyloseq: an R package for reproducible interactive analysis and graphics of microbiome census data. PLoS ONE **8**, e61217 (2013). https://doi.org/10.1371/journal.pone.0061217

21. Schubert, M., Lindgreen, S., Orlando, L.: AdapterRemoval v2: rapid adapter trimming, identification, and read merging. BMC Res. Notes **91**(9), 1–7 (2016). https://doi.org/10.1186/S13104-016-1900-2

22. Zhang, J., Kobert, K., Flouri, T., Stamatakis, A.: PEAR: a fast and accurate Illumina paired-End reAd mergeR. Bioinformatics **30**, 614–620 (2014). https://doi.org/10.1093/bioinformatics/btt593

23. Schmieder, R., Edwards, R.: Quality control and preprocessing of metagenomic datasets. Bioinformatics **27**, 863–864 (2011). https://doi.org/10.1093/bioinformatics/btr026

24. Rognes, T., Flouri, T., Nichols, B., Quince, C., Mahé, F.: VSEARCH: a versatile open source tool for metagenomics. Peer J. **4**, e2584 (2016). https://doi.org/10.7717/PEERJ.2584

25. Cole, J.R., et al.: Ribosomal database project: data and tools for high throughput rRNA analysis. Nucleic Acids Res. **42**, D633–D642 (2014). https://doi.org/10.1093/NAR/GKT1244
26. Benson, D.A., et al.: GenBank. Nucleic Acids Res. **41**, D36–D42 (2013). https://doi.org/10.1093/NAR/GKS1195
27. Tatusova, T.A., Madden, T.L.: BLAST 2 Sequences, a new tool for comparing protein and nucleotide sequences. FEMS Microbiol. Lett. **174**, 247–250 (1999). https://doi.org/10.1111/j.1574-6968.1999.tb13575.x
28. Bengtsson-Palme, J., et al.: Improved software detection and extraction of ITS1 and ITS2 from ribosomal ITS sequences of fungi and other eukaryotes for analysis of environmental sequencing data. Meth. Ecol. Evol. **4**, 914–919 (2013). https://doi.org/10.1111/2041-210X.12073
29. McDonald, D., et al.: The biological observation matrix (BIOM) format or: how I learned to stop worrying and love the ome-ome. Gigascience **1**(1), 2047-217X (2012). https://doi.org/10.1186/2047-217X-1-7
30. Gohl, D.M., et al.: Systematic improvement of amplicon marker gene methods for increased accuracy in microbiome studies. Nat. Biotechnol. **349**(34), 942–949 (2016). https://doi.org/10.1038/nbt.3601
31. Bakker, M.G.: A fungal mock community control for amplicon sequencing experiments. Mol. Ecol. Resour. **18**, 541–556 (2018). https://doi.org/10.1111/1755-0998.12760
32. Bista, I., et al.: Performance of amplicon and shotgun sequencing for accurate biomass estimation in invertebrate community samples. Mol. Ecol. Resour. **18**, 1020–1034 (2018). https://doi.org/10.1111/1755-0998.12888
33. Encyclopedia of Machine Learning: Encycl. Mach. Learn. (2010). https://doi.org/10.1007/978-0-387-30164-8
34. Toju, H., Tanabe, A.S., Yamamoto, S., Sato, H.: High-coverage ITS primers for the DNA-based identification of ascomycetes and basidiomycetes in environmental samples. PLoS ONE **7**, e40863 (2012). https://doi.org/10.1371/JOURNAL.PONE.0040863

Short Papers

CEvADA: Co-Evolution Analysis Data Archive

Neli José da Fonseca Júnior[1,2]([⊠]) [iD], Marcelo Querino Lima Afonso[1] [iD], and Lucas Bleicher[1] [iD]

[1] Department of Biochemistry and Immunology, Institute of Biological Sciences, Federal University of Minas Gerais, Belo Horizonte, MG 31270-901, Brazil
lbleicher@icb.ufmg.br
[2] Cellular Structure and 3D Bioimaging, European Molecular Biology Laboratory, European Bioinformatics Institute, Wellcome Genome Campus, Hinxton CB10 1SA, UK
neli@ebi.ac.uk

Abstract. CEvADA is a database of amino acid coevolution networks aimed to detect specificity determinant and function related sites in protein families. The database was also designed to provide an easy access to protein coevolutionary constraints that can be incorporated in machine learning classification models, just as sequence annotation and structure prediction methods. The data can be accessed for the whole protein family and specific protein sequences. We also provide sequence search and a REST API for programmatic access in the database. The current version of the database contains data related to 6.301 conserved domains and 45 million protein sequences. CeVADA is free and can be accessed at http://bioinfo.icb.ufmg.br/cevada.

Keywords: Amino acid coevolution · Database · Multiple sequence alignments · Proteins

1 Introduction

Amino acid conservation is one of the oldest and most important estimators of structural and functional importance in computational molecular biology, as the lack of mutations accumulated over generations could suggest evolutionary constraints when comparing paralog proteins [6,19,23]. Although homologous sequences commonly share an overall main function, there is also a degree of variance in subsets of these proteins, mainly when comparing proteins after a gene duplication event [5,10]. Therefore, subset-specific conserved residues can also highlight function and structural importance for that individual group of sequences. An example can be seen in Bachega et al. (2009), where the authors could predict and distinguish the sets of residues involved in the iron and manganese superoxide dismutases by using statistical coupling analysis to find the groups of co-occurring residues in the multiple sequence alignment [2,14].

Although there are many public databases related to protein families, there is still a lack of information available regarding functional subfamilies since most

© Springer Nature Switzerland AG 2021
P. F. Stadler et al. (Eds.): BSB 2021, LNBI 13063, pp. 119–124, 2021.
https://doi.org/10.1007/978-3-030-91814-9_11

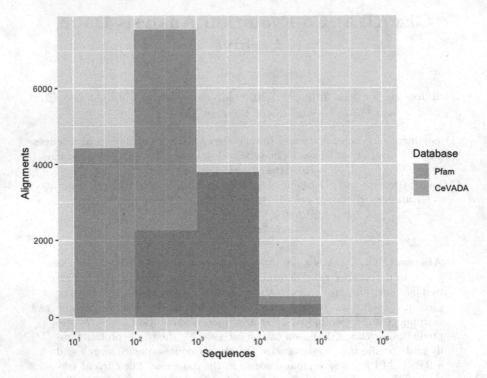

Fig. 1. Coverage of CeVADA in Pfam families in terms of the number of alignments and the number of sequences in each alignment.

of these resources are either interested in classify the entire protein family or single-domain components [17]. Protein function Information in the residue level is even more scarse. At the moment, less than 10% of the entries in UniProt are experimentally characterised, and less than 1% contains functional sites classification [17].

Residue co-variation networks can be used to find function-related residues and make data-driven research possible, as they can highlight patterns present in uncharacterized proteins. This information can be useful in structure-function reviews of protein families [11,13,15,16], rationalizing mutagenesis experiments [3,18] and even in the annotation and characterization of novel proteins [7,9,17].

CEvADA is a database of pre-calculated amino acid coevolution networks for conserved domains deposited in the Pfam [8]. It was designed to solve two common issues in molecular co-evolutionary analysis: 1) provide instant access to the data through the CEvADA REST API and the entry pages, as the calculation of these networks can take a considerable amount of time. 2) provide easy access to patterns from larger protein families, as these software usually have a limit in the size of the input alignment or require access to high-performance computing clusters.

2   Technical Notes

The data was generated using CONAN's standalone package [10] and multiple sequence alignments from Pfam 32.0 (full). Sequences were discarded according to the 80% maximum identity threshold, and coevolution was calculated for residues presenting a positional frequency between 5% and 95%. Two metrics were used to calculate the correlations: the overlap score, computed by the Jaccard similarity coefficient [12]; and the probability of the correlation have occurred by random chance, using a hypergeometric distribution [20]. Two thresholds were applied to remove the weak correlation signals: the minimum negative log of the p-value of 15 and the maximum average Jaccard similarity coefficient of 0.6. The database stores several cuts of the networks in intervals of 0.05 average Jaccard coefficient. At the moment, CeVADA covers 35% of the Pfam families, including all domains containing between 500 and 20.000 sequences after the filtering procedures. Unfortunately, as shown in Fig. 1, most of the full alignments in Pfam 32.0 contain a relatively small number of non-redundant sequences. Therefore these domains are still not included in CeVADA, as this analysis requires a higher sampling to be able to validate the concurrences. However, it is expected that many of those families will eventually be included, given the constant increase in sequence availability from new genome projects.

The CeVADA REST API contains two public endpoints allowing third-party applications to have easy programmatic access to the database. The users can fetch data from a single protein family by passing a Pfam ID or accession code. In addition, it is also possible to retrieve data from a specific protein by providing a Uniprot ID, accession code or sequence. The communication is performed by GET method, and the results are returned in the JSON format. The protein family endpoint gives back the multiple sequence alignment and all the detected residues for the given Pfam domain, numbered according to the positions in the alignment. The protein sequence endpoint returns the co-variation data for all the Pfam domains presented in the given protein sequence. In this case, the positions are numbered according to the given sequence and divided into sets of matches and mismatches.

The website also includes sequence and families entry pages with many data visualizations and cross-references to external resources. The entry pages are powered by internal and external APIs, such as UniprotKb, Gene Ontology and Wikipedia. The protein sequences page includes the description and general information extracted from the UniprotKb and the list of Pfam domains and their regions in the protein. For each domain, there is a list of correlated residues sets. In the protein family pages, the user has access to a general description of the domain extracted from the Wikipedia, GO, and INTERPRO annotations extracted from Pfam. The page also includes views of the networks, positions in the sequences and the co-variation matrices.

The web application was built in Python using the Django framework and the following data visualization packages: visNetwork [1], MSAViewer [22], ProtVista [21] and d3 [4].

122   N. J. da Fonseca Júnior et al.

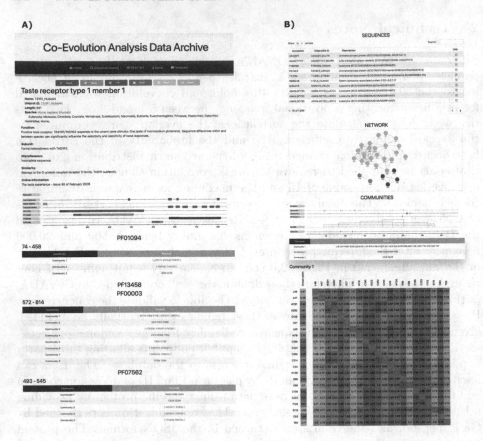

Fig. 2. CeVADA entry pages: A) the protein sequence entry page corresponding to TS1R1_HUMAN. The page is showing all its four conserved domains and their conserved residues (mismatches are displayed in red); B) part of the protein family entry page corresponded to the 7 transmembrane sweet-taste receptor of 3 GCPR (PF00003), including the list of sequences, the coevolution network and the covariation matrices. (Color figure online)

Acknowledgements. Authors thank CEPAD-ICB, UFMG, for hosting CeVADA in the HPC cluster Sagarana.

Funding. This work has been supported by CNPq (grant 457851/2014-7) and CAPES (grant 051/2013).

References

1. Almende, B., Benoit, T., Titouan, R.: Package 'visnetwork'. Netw. Visu. 'vis. js' Lib. Vers. **2**(9) (2019). https://mran.microsoft.com/snapshot/2016-10-12/web/packages/visNetwork/visNetwork.pdf

2. Bachega, J.F.R., et al.: Systematic structural studies of iron superoxide dismutases from human parasites and a statistical coupling analysis of metal binding specificity. Proteins: Struct. Funct. Bioinf. **77**(1), 26–37 (2009). https://doi.org/10.1002/prot.22412

3. Barwinska-Sendra, A., et al.: An evolutionary path to altered cofactor specificity in a metalloenzyme. Nat. Commun. **11**(1), 1–13 (2020). https://doi.org/10.1038/s41467-020-16478-0

4. Bostock, M., Ogievetsky, V., Heer, J.: D^3 data-driven documents. IEEE Trans. Vis. Comput. Graph. **17**(12), 2301–2309 (2011). https://doi.org/10.1109/TVCG.2011.185

5. Chakraborty, A., Chakrabarti, S.: A survey on prediction of specificity-determining sites in proteins. Briefings Bioinf. **16**(1), 71–88 (2015). https://doi.org/10.1093/bib/bbt092

6. Choi, Y., Sims, G.E., Murphy, S., Miller, J.R., Chan, A.P.: Predicting the functional effect of amino acid substitutions and indels. PloS one **7**(10), e46688 (2012). https://doi.org/10.1371/journal.pone.0046688

7. Coitinho, J.B., et al.: Structural and immunological characterization of a new nucleotidyltransferase-like antigen from Paracoccidioides brasiliensis. Mol. Immunol. **112**, 151–162 (2019). https://doi.org/10.1016/j.molimm.2019.04.028

8. El-Gebali, S., et al.: The Pfam protein families database in 2019. Nucleic Acids Res. **47**(D1), D427–D432 (2019). https://doi.org/10.1093/nar/gky995

9. da Fonseca, N.J., Afonso, M.Q.L., de Oliveira, L.C., Bleicher, L.: A new method bridging graph theory and residue co-evolutional networks for specificity determinant positions detection. Bioinformatics **35**(9), 1478–1485 (2019). https://doi.org/10.1093/bioinformatics/bty846

10. Fonseca, N., Afonso, M., Carrijo, L., Bleicher, L.: Conan: a web application to detect specificity determinants and functional sites by amino acids co-variation network analysis. Bioinformatics (2020). https://doi.org/10.1093/bioinformatics/btaa713

11. da Fonseca, N.J., Afonso, M.Q.L., Pedersolli, N.G., de Oliveira, L.C., Andrade, D.S., Bleicher, L.: Sequence, structure and function relationships in flaviviruses as assessed by evolutive aspects of its conserved non-structural protein domains. Biochem. Biophys. Res. Commun. **492**(4), 565–571 (2017). https://doi.org/10.1016/j.bbrc.2017.01.041

12. Jaccard, P.: The distribution of the flora in the alpine zone. 1. New Phytol. **11**(2), 37–50 (1912).https://doi.org/10.1111/j.1469-8137.1912.tb05611.x

13. Lima Afonso, M., de Lima, L., Bleicher, L.: Residue correlation networks in nuclear receptors reflect functional specialization and the formation of the nematode-specific P-box. BMC Genomics **14**(Suppl 6), S1 (2013)

14. Lockless, S.W., Ranganathan, R.: Evolutionarily conserved pathways of energetic connectivity in protein families. Science **286**(5438), 295–299 (1999). https://doi.org/10.1126/science.286.5438.295

15. Oliveira, A., Bleicher, L., Schrago, C.G., Junior, F.P.S.: Conservation analysis and decomposition of residue correlation networks in the phospholipase a2 superfamily (pla2s): Insights into the structure-function relationships of snake venom toxins. Toxicon **146**, 50–60 (2018). https://doi.org/10.1016/j.toxicon.2018.03.013

16. Querino Lima Afonso, M., da Fonseca, N.J., de Oliveira, L.C., Lobo, F.P., Bleicher, L.: Coevolved positions represent key functional properties in the trypsin-like serine proteases protein family. J. Chem. Inf. Model. **60**(2), 1060–1068 (2020). https://doi.org/10.1021/acs.jcim.9b00903

17. Rauer, C., Sen, N., Waman, V.P., Abbasian, M., Orengo, C.A.: Computational approaches to predict protein functional families and functional sites. Curr. Opin. Struct. Biol. **70**, 108–122 (2021). https://doi.org/10.1016/j.sbi.2021.05.012
18. Rios-Anjos, R.M., de Lima Camandona, V., Bleicher, L., Ferreira-Junior, J.R.: Structural and functional mapping of Rtg2p determinants involved in retrograde signaling and aging of Saccharomyces cerevisiae. PloS One **12**(5) (2017). https://doi.org/10.1371/journal.pone.0177090
19. Taylor, W.R.: The classification of amino acid conservation. J. Theor. Biol. **119**(2), 205–218 (1986). https://doi.org/10.1016/s0022-5193(86)80075-3
20. Tumminello, M., Miccichè, S., Lillo, F., Piilo, J., Mantegna, R.N.: Statistically validated networks in bipartite complex systems. PLoS ONE **6**(3) (2011). https://doi.org/10.1371/journal.pone.0017994
21. Watkins, X., Garcia, L.J., Pundir, S., Martin, M.J., Consortium, U.: Protvista: visualization of protein sequence annotations. Bioinformatics **33**(13), 2040–2041 (2017). https://doi.org/10.1093/bioinformatics/btx120
22. Yachdav, G., et al.: Msaviewer: interactive javascript visualization of multiple sequence alignments. Bioinformatics **32**(22), 3501–3503 (2016). https://doi.org/10.1093/bioinformatics/btw474
23. Zuckerkandl, E., Pauling, L.: Evolutionary divergence and convergence in proteins. In: Evolving Genes and Proteins, pp. 97–166. Elsevier (1965). https://doi.org/10.1016/B978-1-4832-2734-4.50017-6

FluxPRT: An Adaptable and Extensible Proteomics LIMS

Elizabeth Regina Alfaro-Espinoza[1,2,4(✉)] , Lucas Ferreira Paiva[3] ,
Alessandra C. Faria-Campos[2,5] , Maria Cristina Baracat-Pereira[4] ,
and Sérgio Vale Aguiar Campos[2]

[1] Programa de Pós -Graduação em Bioinformática, Instituto de Ciências Biológicas,
Universidade Federal de Minas Gerais, Belo Horizonte, Brazil
elizaespinoza@ufmg.br
[2] Laboratório de Universalização de Acesso, Departamento de Ciência da
Computação, Universidade Federal de Minas Gerais, Belo Horizonte, Brazil
{alessa,scampos}@dcc.ufmg.br
[3] Programa de Pós-Graduação em Ciência da Computação,
Departamento de Informática, Universidade Federal de Viçosa, Viçosa, Brazil
lucas.paiva@ufv.br
[4] Laboratório de Proteômica e Bioquímica de Proteínas,
Departamento de Bioquímica e Biologia Molecular, Universidade Federal de Viçosa,
Viçosa, Brazil
baracat@ufv.br
[5] Laboratório de Bioengenharia - Instituto Nacional de Metrologia,
Qualidade e Tecnologia, Viçosa, Brazil

Abstract. Proteomics is a fundamental research area that focuses
on large-scale protein analysis. Data generated from experimental
approaches in proteomics is becoming increasingly complex to orga-
nize and manage. Laboratory Information Management Systems (LIMS)
allow this data to be managed in a flexible and efficient way. However,
most LIMS have a high cost and are directed at specific needs. This
work introduces FluxPRT, a comprehensive and adaptable proteomics
workflow-based LIMS supported by an interactive web page guide aimed
at assisting novice researchers in a proteomics lab. FluxPRT is available
at http://www.flux2.luar.dcc.ufmg.br and provides an important tool to
assist preotemics research.

Keywords: Workflow-based data · Proteomics laboratory · LIMS and
Information management · Data collecting

1   Introduction

Proteomics experiments can provide invaluable scientific information that may
lead to important scientific discoveries. Modern proteomic methods allow these

Supported by Capes. Fellowship: 88887.517813/2020-00.

experiments to be performed very efficiently. A proteomics workflow involves several steps, such as Orthogonal Protein Separations (Electrophoresis and Chromatographies); Mass Spectrometry (MS); and Proteomics Informatics (MS Informatics and Quantification) [7]. Therefore, proteomics experiments can be very complex and tracking all data related to an experiment can be challenging. Computer systems can be an important aid in this matter. One type of such systems is Laboratory Information Management Systems (LIMS).

LIMS are complex computer systems used to store and manage laboratory data. Their main focus is guaranteeing the quality of the processes and ensuring that results are produced consistently and reliably. They control the entire data life cycle, from sample preparation to result analysis [5]. Several LIMS are currently available [8]. However, those are usually too complex and expensive, and most are specialized in specific areas, making proteomics LIMS a desirable, but hard to find tool.

LIMS, pipelines and analysis tools for mass spectrometry [9], quantitative proteomics [10], and metaproteomics [2], are some of the proteomics management tools available today. However, these solutions only control a subset of the processes involved in a proteomics workflow and lack information to assist inexperienced users in executing proteomics experiments.

In this work, we used the Flux LIMS [4], a workflow-based flexible LIMS designed to manage laboratory data efficiently and reliably, to address this problem. Flux has been used as the basis for designing FluxPRT, a proteomics LIMS that is flexible and powerful for managing proteomics experimental data. FluxPRT registers all actions, from sample collection through protein identification. Additionally, we present an interactive guide that can be accessed as a standalone web page to help new researchers with proteomics information.

2 Proteomics Lab Operations

Proteomics is the complex process of identifying and quantifying proteins expressed under different conditions or phases of a cell or organism's life. It is performed on a large-scale and might vary considerably depending on the protein source and the technology available [11].

Proteomic Analysis includes obtaining protein source material, extraction and purification of proteins by chromatography and/or electrophoresis, analysis by mass spectrometry (MS) to characterize proteins using their mass to charge ratios (m/z) and relative abundances and identification of those by matching the values obtained in MS to spectra in a database.

Three different approaches can be used in proteomics research. Partially digested proteins and big protein fragments can be analyzed in MS using *middle-down* proteomics [6]. It offers the benefit of giving higher proteome coverage, including identification of splice variants and additional isoforms. The *Bottom-up* or *shotgun* approach relies on isolating peptides for MS analysis. It has the ability to be able to resolve the majority of full proteomes [3]. However, it lacks the sensitivity required to detect proteoforms and post-translational modifications (PTMs). In the *top-down* strategy, complete proteins are introduced for

MS analysis, which represents a highly efficient strategy to assess PTMs and isoforms [1].

3   Methodology

3.1   Proteomics Workflow Construction and the Flux LIMS

Proteomic Analysis in a laboratory has been represented in Flux using a work-flow. The workflow has been modeled using the draw.io tool (https://www.draw.io/) after analyzing the demands of a proteomics laboratory. The sequence of activities that compose the workflow represents these demands. Each activity represents actions in a proteomics lab and has a set of attributes, which are defined inside the activities. Each attribute has a type (ex: *'string'*, *'integer'*, *'picture'*, *'register'*, etc.), and activities can have precedence connections.

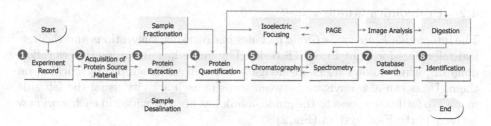

Fig. 1. The FluxPRT workflow.

The modeled workflow was used to construct an XPDL workflow using a built-in Workflow Editor in Flux, which was refined in collaboration with experts from LPBP, a laboratory at UFV (Brazil). The workflow, named **FluxPRT** workflow, was uploaded and tested in Flux. Flux is a workflow-based LIMS that uses Java technology, with MySQL as database server and Apache Tomcat as web server. Flux web interface is accessible via the major web browsers and different workflow files can be uploaded, resulting in a more flexible system.

3.2   Proteomics Guide

To assist inexperienced users in using FluxPRT, a help is provided in the system as well as in a proteomics guide. The *Help* button in the system provides access to information about the Flux functions and features, along with a user manual. Additionally, FluxPRT-specific documentation was created, through an interactive guide with descriptions of the activities represented in the FluxPRT workflow.

The interactive guide is a GitHub hosted web page built using HTML, JavaScript and CSS. It was created using research and review articles as well as proteomics protocol books. The web page was developed to provide information for screen reading technologies, with the goal of making it accessible to blind people. Furthermore, the entire guide's content was organized into an ebook and made available on the website, allowing users to access information offline. The visual information from the ebook was also examined in a color blindness simulator.

4   Results and Discussion

4.1   FluxPRT Workflow

FluxPRT proteomics workflow has 14 activities (Fig. 1), which represent proteomic analysis from sample preparation to protein identification including MS and bioinformatics analysis. Moreover, it allows experiment tracking and protocol management.

4.2   Proteomics Guide

The interactive *Proteomics Guide* includes instructions, suggestions and theoretical references for eight FluxPRT tasks. The guide has an interactive workflow diagram, which allows users to navigate between tasks simply by clicking on them. Users can also navigate between steps in each activity using the left side menu. To facilitate access to the guide, a link has been provided in each workflow activity in the Flux system (Fig. 2).

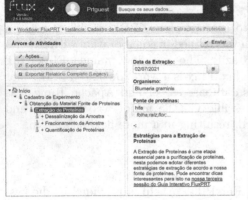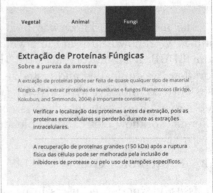

Fig. 2. FluxPRT interface and the Proteomics guide access.

4.3   FluxPRT Interface

FluxPRT has an easy-to-use interface where activities that represent proteomic analysis steps are grouped in an *Activity tree*. Activities that have already been executed are represented by test tube icons in a different color from those for activities that are available to be executed (Fig. 2). The system guides the users through the entire process, informing them of which activities are available for execution. It is also possible to modify, remove, or disable an activity and generate reports in pdf.

It should be noticed that FluxPRT is built in such a way that users can register data according to the approach used (Fig. 3), rather than being limited to a particular method of performing an experiment. Furthermore, depending on the aim of the experiment, each activity can be performed more than once and can be tracked in FluxPRT. For example, in the Chromatography step, a first exploration can be executed and another HPLC-type chromatography step can be conducted.

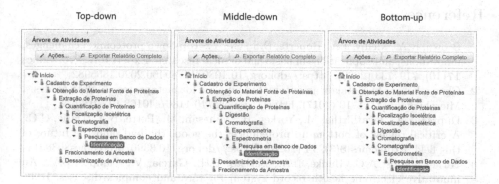

Top-down   Middle-down   Bottom-up

Fig. 3. Examples of three potential experiment flows based on *top-down*, *middle-down* and *bottom-up* proteomics approaches.

With these functionalities, the FluxPRT system intends to be a support tool for laboratory coordinators and technicians, keeping an accurate record of the development of various projects inside the laboratory and regulating the level of access for each member, enabling data traceability.

5   Concluding Remarks

This study presents the FluxPRT, a proteomics workflow based-LIMS that is complemented by an interactive guide. It offers several functionalities, including sample management, report generation and the possibility to repeat experiment steps and adjust the process to different approaches, as illustrated by the integration of three entirely different proteomics approaches.

This ongoing work seeks to assist proteomics researchers in organizing experiments and maximizing the learning of members with less expertise. In the future, the workflow will be adapted to the MIAPE standards, the interactive guide will be completed to cover other workflow activities and FluxPRT suitability must be tested more intensively in a proteomics laboratory. The current version of FluxPRT is only in Portuguese, but the system is currently being translated to English.

6   Availability

FluxPRT is available at http://www.flux2.luar.dcc.ufmg.br using the user guest2021 and password gu3st, and the documentation website can be accessed at https://fluxprt.github.io. Currently the FluxPRT system is in user tests, if you are interested in applying this system to your laboratory, please contact the corresponding author.

References

1. Brown, K.A., Melby, J.A., Roberts, D.S., Ge, Y.: Top-down proteomics: challenges, innovations, and applications in basic and clinical research. Expert Rev. Proteomics **17**(10), 719–733 (2020). https://doi.org/10.1080/14789450.2020.1855982
2. Cheng, K., et al.: Metalab: an automated pipeline for metaproteomic data analysis. Microbiome **5**(1), 1–10 (2017). https://doi.org/10.1186/s40168-017-0375-2
3. Dupree, E.J., Jayathirtha, M., Yorkey, H., Mihasan, M., Petre, B.A., Darie, C.C.: A critical review of bottom-up proteomics: the good, the bad, and the future of this field. Proteomes **8**(3), 14 (2020). https://doi.org/10.3390/proteomes8030014
4. Faria-Campos, A.C., Hanke, L.A., Batista, P.H., Garcia, V., Campos, S.V.: An innovative electronic health record system for rare and complex diseases. BMC Bioinform. **16**(19), 1–8 (2015). https://doi.org/10.1186/1471-2105-16-S19-S4
5. Hinton, M.: LIMS in the manufacturing environment. Lab. Autom. Inf. Manage. **31**(2), 109–113 (1995). https://doi.org/10.1016/1381-141X(95)80027-Z
6. Lermyte, F., Tsybin, Y.O., O'Connor, P.B., Loo, J.A.: Top or middle? up or down? toward a standard lexicon for protein top-down and allied mass spectrometry approaches. J. Am. Soc. Mass Spectrometry **30**(7), 1149–1157 (2019). https://doi.org/10.1007/s13361-019-02201-x
7. Martínez-Bartolomé, S., Binz, P.A., Albar, J.P.: The minimal information about a proteomics experiment (MIAPE) from the proteomics standards initiative. In: Plant Proteomics, pp. 765–780. Springer (2014). https://doi.org/10.1007/978-1-62703-631-3_53
8. Melo, A., Faria-Campos, A., DeLaat, D.M., Keller, R., Abreu, V., Campos, S.: Sigla: an adaptable LIMS for multiple laboratories. BMC Genomics **11**(5), 1–8 (2010). https://doi.org/10.1186/1471-2164-11-S5-S8
9. Vizcaíno, J.A., et al.: Proteomexchange provides globally coordinated proteomics data submission and dissemination. Nature Biotechnol. **32**(3), 223–226 (2014). https://doi.org/10.1038/nbt.2839
10. Wen, B., Zhou, R., Feng, Q., Wang, Q., Wang, J., Liu, S.: Iquant: an automated pipeline for quantitative proteomics based upon isobaric tags. Proteomics **14**(20), 2280–2285 (2014). https://doi.org/10.1002/pmic.201300361
11. Wilkins, M.R., et al.: Information management for proteomics: a perspective. Expert Rev. Proteomics **5**(5), 663–678 (2008). https://doi.org/10.1586/14789450.5.5.663

MathPIP: Classification of Proinflammatory Peptides Using Mathematical Descriptors

João Pedro Uchôa Cavalcante[1], Anderson Cardoso Gonçalves[1],
Robson Parmezan Bonidia[1(✉)] [iD], Danilo Sipoli Sanches[2],
and André Carlos Ponce de Leon Ferreira de Carvalho[1]

[1] Institute of Mathematics and Computer Sciences, University of São Paulo - USP,
São Carlos 13566 -590, Brazil
[2] Bioinformatics Graduate Program (PPGBIOINFO), UTFPR,
Federal University of Technology-Paraná, Campus Cornélio Procópio,
Cornélio Procópio 86300-000, Brazil

Abstract. Proinflammatory peptide (PIP) is a relevant part of the inflammatory response, often the first response of our immune system to strange bodies, i.e., inflammatory-inducing infection, such as COVID-19. Thus, it is essential to have reliable ways to classify and analyze new instances of PIPs. Machine learning (ML) models have been widely employed for the classification of biological sequences, being the basis for most studies in extensive databases of biological information. Most ML algorithms have difficulty to directly deal with these sequences. Thereby, relevant features are extracted from these sequences, making feature extraction one of the key steps in the application of ML algorithms to biological data. Different features have been proposed, many of them based on prior knowledge, such as molecular structures. However, many biological sequences publicly available do not come with prior knowledge. To deal with this limitation, we propose to investigate the use of mathematical descriptors to extract features from PIP sequences. To assess how relevant are the features extracted using mathematical descriptors, we run experiments where we apply three ML algorithms. In these experiments, we obtained a predictive accuracy of 0.7034, which is on par with current PIP classifiers.

Keywords: Feature extraction · Biological sequences · Mathematical descriptors · Machine learning

1 Background

The inflammatory response is often the primary defense mechanism that our bodies use to fight against infections caused by pathogens or other agents [5], such as the pro-inflammatory feedback loop caused by the pathogenesis of the

---

J. P. U. Cavalcante, A. C. Gonçalves and R. P. Bonidia—The authors wish it to be known that, in their opinion, the first three authors should be regarded as Joint First Authors.

P. F. Stadler et al. (Eds.): BSB 2021, LNBI 13063, pp. 131–136, 2021.
https://doi.org/10.1007/978-3-030-91814-9_13

COVID-19 [1]. According to [4], the identification of PIPs is an important topic in immunoinformatics and computational biology, where further studies on these mechanisms are needed, as well as tools to identify them, such as ML pipelines. According to [5], the current methods provide relevant results but with room for further improvement.

Furthermore, existing methods use only conventional descriptors for feature extraction, e.g., ProInflam [3] (composition-based features, physicochemical properties, and motif-based features), PIP-EL [4] (composition-based features, composition-transition-distribution, amino acid index, and physicochemical properties), and ProIn-Fuse [5] (eight types of encoding schemes, e.g. kmer-pr, kmer-ac, and binary). Considering this, we propose a new way to classify proinflammatory peptides (PIP), using and expanding the MathFeature [2] package, a tool that provides multiple mathematical feature descriptors (e.g., Fourier, entropy, numerical mapping, graphs) to numerically represent biological sequences through feature engineering. This process is a fundamental step for ML models applied to biological data. Therefore, we assume the following hypothesis:

– **Hypothesis:** ML models using only mathematical feature descriptors can be as robust as existing models for PIP classification.

Finally, our best model, called MathPIP, present an Accuracy (ACC) of 0.7034, which is higher than some existing studies, e.g., ProInflam [3] (ACC: 0.6280) and PIP-EL [4] (ACC: 0.6490).

2   Materials and Methods

This section will cover our methodological procedures into three stages, them being: Dataset selection, Feature engineering, and Experimental setting.

2.1   Data Selection

We have used a benchmark dataset provided by [5], which was also applied by the author to compare the following tools: ProInflam [3], PIP-EL [4], and PoIn-fuse [5]. We also followed the original division of the dataset with a ratio of 8:2, that is, training with 607 PIPs and 1098 non-PIPs, and testing with 134 PIPs and 156 non-PIPs. The class imbalance between PIP and non-PIP is 3:5 and 9:10, in the train and test, respectively.

2.2   Feature Engineering

A conventional way of extracting features to classify biological sequences is using alignment techniques, which means searching databases for similar known sequences or using conventional descriptors (alignment-free) such as k-mer, amino acid composition, and physicochemical features. These techniques achieve relevant results, as shown in [3–5]. Nevertheless, none of these studies explore mathematical descriptors such as numerical mapping, Fourier, entropy, and graphs.

Thereby, through MathFeature, we can receive a biological sequence, translating it to a Discrete Intermediary State (DIS), and applying math/informational techniques to extract features from the DIS, as shown in Fig. 1.

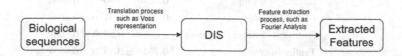

Fig. 1. Feature engineering pipeline.

Finally, for this study, we use the following descriptors: Numerical Mappings [6] (Accumulated Amino Acid Frequency, Integer, and EIIP), Fourier Series [6, 7] with integer mapping, entropy [6,8] (Shannon and Tsallis), and Complex Networks [6].

2.3 Experimental Setting

We chose the python programming language and its libraries, e.g., pandas, xgboost, sklearn, numpy, math, random, and MathFeature. We use the XGBoost (XGB), Random Forest (RF), and Support Vector Machine (SVM) classifiers. The metrics applied were Accuracy score (ACC), Matthews Correlation Coefficient (MCC), and Area Under the ROC Curve (AUC).

3 Results and Discussion

Each model presented in this section consists of a combination of ML classifiers, DIS, and feature extraction descriptors. All models are trained and validated in two ways: (1) the k-fold method (k = 10) and (2) the classical train-test method. The k-fold method consists of splitting the dataset into k parts with an equal number of instances. Then k-1 parts are used to train, and the last one is used to validate. The process of training and validating is repeated k times generating k models. Based on this, we calculate the mean of the scores and assign them as being the ACC, MCC, and AUC of a given model, as shown in Table 1. This validation method is relevant because we are randomly simulating a draw from an unknown distribution of data, and validating it with another random draw, which could help us get closer to the real world.

Thereby, our results show that the best mathematical descriptors are Accumulated AA Frequency with XGB (ACC: 0.7589, MCC: 0.4504, and AUC: 0.6946), following by Shannon Entropy-RF (ACC: 0.7536, MCC: 0.4363, and AUC: 0.6879) and Complex Networks-XGB (ACC: 0.7519, MCC: 0.4327, and AUC: 0.6876). The cross-validation (10 fold) results were robust in almost all models, sometimes by a considerable margin, in the scores. As mentioned above, the results of Table 1 are the means of various combinations of the dataset divided into different training and validation partitions, and some of these combinations had class imbalances that, paired with model bias, likely generated the scores

Table 1. Performance on training dataset using tenfold cross-validation

Classifier	Descriptor	ACC	MCC	AUC
RF	Accumulated AA Frequency	**0.7589**	**0.4499**	**0.6960**
	Integer Mappings	0.7419	0.4084	0.6803
	EIIP Mapping	0.7291	0.3750	0.6593
	Fourier + Integer Mapping	0.6862	0.2605	0.6069
	Tsallis Entropy	0.7513	0.4329	0.6857
	Shannon Entropy	0.7536	0.4363	0.6879
	Complex Networks	0.7490	0.4247	0.6811
XGB	Accumulated AA Frequency	**0.7589**	**0.4504**	**0.6946**
	Integer Mappings	0.7414	0.4063	0.6772
	EIIP Mapping	0.7132	0.3312	0.6359
	Fourier + Integer Mapping	0.6862	0.2605	0.6069
	Tsallis Entropy	0.7472	0.4206	0.6814
	Shannon Entropy	0.7495	0.4246	0.6835
	Complex Networks	0.7519	0.4327	0.6876
SVM	Accumulated AA Frequency	0.7249	0.3626	0.6463
	Integer Mappings	0.7079	0.3153	0.6267
	EIIP Mapping	0.6880	0.2520	0.5923
	Fourier + Integer Mapping	0.6956	0.2791	0.6082
	Tsallis Entropy	0.6991	0.2898	0.6160
	Shannon Entropy	0.6968	0.2833	0.6135
	Complex Networks	**0.7255**	**0.3637**	**0.6555**

with the highest margin. This does not constitute a problem, since a drop in the metrics of a given model is expected when applied to real problems, as exemplified in the next experiment. That is, we also test all models using the train-test data splitting method, the most common way to train ML classifiers, that for this specific dataset was used by other studies in the literature. We used the same proportions on train and test dataset (8:2), based on [5], as shown in Table 2.

Again, in testing dataset (simulating samples not seen), the best results were of Accumulated AA Frequency with XGB (ACC: 0.7034, MCC: 0.4085, and AUC: 0.6434), following by Complex Networks-XGB (ACC: 0.6931, MCC: 0.3838, and AUC: 0.6977) and Accumulated AA Frequency-RF (ACC: 0.6931, MCC: 0.3847, and AUC: 0.6495). So, the best model of this paper (Accumulated AA Frequency with XGB), here called MathPIP, was compared with existing studies, using the same testing dataset (see Table 3).

MathPIP (XGB + Accumulated AA Frequency) was more accurate than ProInflam and PIP-EL by a robust margin, if you take in count the MCC, e.g., 0.1445 (14.45%) and 0.1095 (10.95%), respectively. Our results were worse when compared to ProIn-fuse, with a difference of 4.26% in ACC. Nevertheless, ProIn-fuse uses a fusion of eight feature descriptors that can explain its performance. This hybrid approach is not tested in our study, which could possibly contribute

Table 2. Performance on test dataset

Classifier	Descriptor	ACC	MCC	AUC
RF	Accumulated AA Frequency	**0.6931**	**0.3847**	**0.6495**
	Integer Mappings	0.6483	0.2887	0.6496
	EIIP Mapping	0.6207	0.2321	0.6244
	Fourier + Integer Mapping	0.5759	0.1318	0.5567
	Tsallis Entropy	0.6690	0.3335	0.6760
	Shannon Entropy	0.6690	0.3317	0.6760
	Complex Networks	0.6724	0.3441	0.6613
XGB	Accumulated AA Frequency	**0.7034**	**0.4085**	**0.6434**
	Integer Mappings	0.6724	0.3412	0.6768
	EIIP Mapping	0.6276	0.2519	0.6363
	Fourier + Integer Mapping	0.5793	0.1401	0.5510
	Tsallis Entropy	0.6759	0.3468	0.6843
	Shannon Entropy	0.6724	0.3398	0.6449
	Complex Networks	0.6931	0.3838	0.6977
SVM	Accumulated AA Frequency	**0.6586**	**0.3268**	**0.6142**
	Integer Mappings	0.6172	0.2304	0.6014
	EIIP Mapping	0.6000	0.1949	0.6353
	Fourier + Integer Mapping	0.5862	0.1568	0.5646
	Tsallis Entropy	0.5621	0.1026	0.4945
	Shannon Entropy	0.5759	0.1323	0.5174
	Complex Networks	0.6345	0.2614	0.6773

Table 3. Comparison with existing studies

Method	ACC	MCC
ProInflam	0.6280	0.2640
PIP-EL	0.6490	0.2990
ProIn-fuse	**0.7460**	**0.4880**
MathPIP	**0.7034**	**0.4085**

to the model's performance, however, our premise was to use well-known ML models combined with a mathematical extraction method to test our hypothesis. Moreover, to the best of our knowledge, our proposal is the first to use mathematical descriptors to classify PIPs, indicating possible contributions to hybrid approaches, as presented by existing studies. Finally, our findings indicate that our initial study hypothesis, *ML models using mathematical feature descriptors can be as robust as existing models for PIP classification,* is true since our tool (MathPIP) showed superior (ProInflam and PIP-EL) and competitive (ProIn-fuse) predictive performance.

4   Conclusion

PIP is a relevant part of the inflammatory response, in which reliable ways to classify and analyze new instances are needed. Thereby, considering that the existing methods use only conventional descriptors for feature extraction, we propose a new way to classify PIPs, expanding the MathFeature package, a tool that provides multiple mathematical feature descriptors. We assessed our hypothesis using a benchmark dataset with 1995 sequences (1705 for training and 290 for testing). Furthermore, we evaluate seven mathematical descriptors with three ML classifiers (RF, XGB, and SVM), using three metrics (ACC, MCC, and AUC). The best model, here call MathPIP, reached ACC, MCC, and AUC of 0.7034, 0.4085, and 0.6434, respectively. MathPIP compared to the other existing studies, showed superior and competitive predictive performance, reinforcing that our hypothesis is valid. Finally, these results are important because they suggest new possibilities for a more generalized feature engineering process to classify PIP sequences with ML models.

Acknowledgments. The authors would like to thank ICMC-USP and Coordenação de Aperfeiçoamento de Pessoal de Nível Superior (CAPES) for the financial support given to this research.

References

1. Tay, M.Z., Poh, C.M., Rénia, L., et al.: The trinity of COVID-19: immunity, inflammation and intervention. Nat. Rev. Immunol. **20**, 363–374 (2020). https://doi.org/10.1038/s41577-020-0311-8
2. Bonidia, R.P., Sanches, D.S., de Carvalho, A.C.: Mathfeature: feature extraction package for biological sequences based on mathematical descriptors. bioRxiv (2020)
3. Gupta, S., Madhu, M.K., Sharma, A.K., Sharma, V.K.: ProInfam: a webserver for the prediction of proinflammatory antigenicity of peptides and proteins. J. Transl. Med. **14**(1), 178 (2016)
4. Manavalan, B., Shin, T.H., Kim, M.O., Lee, G.: PIP-EL: a new ensemble learning method for improved proinflammatory peptide predictions. Front. Immunol. **9**, 1783 (2018)
5. Khatun, M.S., Hasan, M.M., Shoombuatong, W., Kurata, H.: ProIn-Fuse: improved and robust prediction of proinflammatory peptides by fusing of multiple feature representations. J. Comput.-Aided Molecular Des. **34**(12), 1229–1236 (2020). https://doi.org/10.1007/s10822-020-00343-9
6. Bonidia, R.P.: Feature extraction approaches for biological sequences: a comparative study of mathematical features. Brief. Bioinform. **22**(5), bbab011 (2021)
7. Cochran, W.T.: What is the fast Fourier transform? Proc. IEEE, **55**(10), 1664–1674 (1967)
8. Machado, J.T., Costa, A.C., Quelhas, M.D.: Shannon, Rényie and Tsallis entropy analysis of DNA using phase plane. Nonlinear Anal. Real World Appl. **12**(6), 3135–3144 (2011)
9. Costa, L.D.F., Rodrigues, F.A., Cristino, A.S.: Complex networks: the key to systems biology. Gene. Molecular Biol. **31**(3), 591–601 (2008)

Metagenomic Insights of the Microbial Community from a Polluted River in Brazil 2020

Carolina O. P. Gil[1](✉), Larissa Macedo Pinto[2], Flavio F. Nobre[1],
Cristiane Thompson[2], Fabiano Thompson[2], and Diogo Antonio Tschoke[1,2]

[1] COPPE/PEB - UFRJ, Rio de Janeiro, Brasil
carol.pgil@peb.ufrj.br
[2] Graduate Program in Genetics, Federal University of Rio de Janeiro – UFRJ,
Rio de Janeiro, Brazil

Abstract. The water in the Metropolitan Region of Rio de Janeiro and in some municipalities in the Baixada Fluminense comes from the hydrological basin of the Guandu River (GR) and its potability is guaranteed by the Water Treatment Station of the Companhia Estadual de Águas e Esgotos. Along its route to the CEDAE dams, the GR suffers urban influences, being heavily impacted by receiving in natura effluents. To check the quality of the water in the GR, daily monitoring is carried out throughout its distribution network, including bacteriology. However, so far there is no metagenomics work to know what are the other microorganisms that exist in the GR that can cause diseases. This work aims to perform a metagenomic analysis of the GR to assess its diversity. Samples distributed in the catchment area of CEDAE and drinking water were collected, submitted to DNA sequencing using Illumina. Quality control of Qpherd >30 sequences and joining of paired-end sequences forward and reverse with the Prinseq program. 203,951,644 sequences were obtained. The bacterial diversity index analysis did not show significant differences among the samples. The most abundant class was Betaproteobacteria. The cluster analysis showed to be significant for the drinking water sample to be grouped together with the raw water sample. The PCA-Biplot showed three clusters and which variables differentiate the samples, some genera having great contributions such as: *Staphylococcus, Chthoniobacter* and *Riemerella*.

Keyword: Metagenomic · Diversity indexes · River

1 Introduction

The water in the Metropolitan Region of Rio de Janeiro and in some municipalities in the Baixada Fluminense comes from the hydrological basin of the Guandu River, and its potability is guaranteed by the Water Treatment Station (WTS) of the State Water and Sewage Company (CEDAE), specifically WTS Guandu, which supplies eight municipalities and has a flow capacity to supply a population of nine million people, being considered the largest WTS in the world [1].

© Springer Nature Switzerland AG 2021
P. F. Stadler et al. (Eds.): BSB 2021, LNBI 13063, pp. 137–144, 2021.
https://doi.org/10.1007/978-3-030-91814-9_14

Along its route to the CEDAE dams, the Guandu River suffers urban influences, mainly in the municipalities of Japeri, Engenheiro Pedreira and Seropédica, being greatly impacted by receiving in natura effluents. Within the municipality of Rio de Janeiro, it is the main and only river with the capacity to supply water to the population of more than six million [2], according to the last IBGE census carried out in 2010 [3].

To check the quality of its water, daily monitoring is carried out along its distribution network and at the treatment outlets of the treatment stations and water sources. Analysis of taste and odor, presence of cyanotoxins, geosmin/MIB (2 Methyl-Isoborneol), bacteriology (thermotolerant coliforms and *Escherichia coli*) and physical-chemical analysis such as: pH, turbidity, conductivity, fluoride and free residual chlorine are performed [4].

However, despite the daily monitoring, so far there is no metagenomics work, to know what are the other microorganisms that exist in the Guandu River. Which is an additional concerning, since some of these microorganisms possible can cause diseases, are resistant to antibiotics and/or heavy metals. Metagenomic analysis makes this knowledge possible, in addition to being important in environmental management, as it can provide necessary information to properly manage this aquatic ecosystem to better understand the environment and effectively allocate investments and actions [2].

Thus, metagenomics is the best way to identify the diversity of microorganisms in the environment, especially the more complex ones, as it is a culture-independent technique [5]. It uses nucleotide sequencing analysis as an approach, a powerful tool to compare and explore the ecology, metabolism and evolution of community profiles of environmental microbiomes, the digestive system microbiome of humans and animals [6, 7]. Therefore, this work aims to perform metagenomic analysis of the Guandu River to assess its diversity and abundance of disease-causing microorganisms and bacteria.

2 Materials and Methods

The procedures described in this section were adopted in all the collections carried out in the Guandu River in January, February and March 2020, where eight samples were collected (Table 1).

Table 1. Identification of codes with dates of collections and location

Samples	Collection date	Location	Latitude	Longitude
Capt_1_Fev_20	02/01/2020	Water catchment channel	22°48′24″S	43°37′33″W
Capt_2_Fev_20	02/01/2020	Water catchment channel	22°48′24″S	43°37′33″W
Guandu_Jan_20	01/17/2020	Guandu River	22°47′33.2″S	43°37′36.5″W
Capt_Jan_20	01/17/2020	Water catchment channel	22°48′24″S	43°37′33″W
Capt_1_Mar_20	03/09/2020	Near the water catchment channel	22°48′24″S	43°37′33″W
Capt_2_Mar_20	03/09/2020	Near the water catchment channel	22°48′24″S	43°37′33″W
Pot_1_Mar_20	03/09/2020	Residence	22°51′21.0″S	43°37′33″W
Pot_2_Mar_20	03/09/2020	Residence	22°50′21.0″S	43°36′32.0″W

Genomic DNA was extracted with the NucleoSpin tissue kit (Macherey-Nagel GmbH & Co. KG) which was used for 150-bp paired-end library preparation with Nextera XT DNA Sample Preparation Kit and sequencing on the NextSeq platform (Senai CETIQT Platform) [8]. The size distribution of the libraries was evaluated using a 2100 Bioanalyzer and a High-Sensitivity DNA kit (Agilent, Santa Clara, CA, USA). A 7500 Real Time PCR machine (Applied Biosystems, Foster City, CA, USA) and a KAPA Library Quantification kit (KapaBiosystems, Wilmington, MA, USA) were used for the quantification of the libraries. The sequences obtained were preprocessed with PRINSEQ software to remove reads smaller than 35 bp and low-score sequences (Phred 30) [9]. Sequence reads were assembled using A5-Miseq software [10] with default parameters.

The exploration of the taxonomic and functional diversity of the microbial community present in the samples was carried out from results obtained with the BlastN local alignment program [11] of the metagenomic sequences against the non-redundant nucleotide database (NT-GenBank) of the NCBI, and the non-redundant protein database (NR-GenBank) from the NCBI, by the Diamond program [12] respectively. The similarity results were analyzed by the MEGAN 6 program ("Metagenome Analyzer") [13] to perform the sequence binning.

All statistical analysis were performed using the R programming language version 4.0.1. Diversity analyses were performed using the Shannon and Simpson indexes and the evenness using the Pielou index [14]. All indexes were calculated using the Vegan package [15].

In addition to the diversity analysis, Hierarchical Cluster analysis was performed, to verify how the samples are grouped and PCA (Principal Component Analysis - Biplot) analysis, which aims to verify if the sample components are similar or not, and which variables are important to discriminate one sample from each other.

3   Results

A total of 203,951,644 sequences were obtained from the eight samples collected (Table 2). In this table it is possible to see the number of sequences that were submitted, and the number of sequences obtained after the Prinseq filtering.

Table 2. Number of sequences obtained from Prinseq and sample names.

ID sample	# Raw data (Read PE – 2×150)	# Reads Post-Prinseq/Pear
Capt1 Fev_20	27459851	27215351
Capt2 Fev_20	39078131	38768421
Capt2 Mar_20	13662840	13057525
Capt1 Mar_20	45459743	44917613
Pot1 Mar_20	68619306	67476603
Pot2 Mar_20	922502	870349
Guandu Jan_20	6398071	6502029
Capt Jan_20	5222485	5143753

Concerning the Bacteria Domain 217 Classes were found, among the three most abundant we found: Betaproteobacteria with an average of 35.51% (of the identified reads), which was expected because the most abundant species belong to this class that are members of freshwater bacterioplankton [16, 17], followed by Actinobacteria with 13.25% (of the reads). Four classes presented relative abundance from 3% to 1% of the reads (Opitutae, unclassified Cloroflexi, Deltaproteobacteria and Flavobacteria) while the others with <1% (Fig. 1).

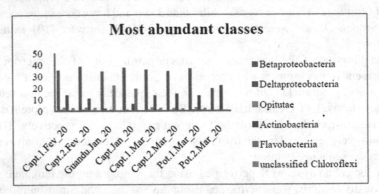

Fig. 1. Most abundant classes of bacteria. Relative abundance of the most frequent classes found in the metagenomic study.

A total of 3097 bacterial genera were found, and from these data the Shannon, Simpson and Pielou diversity indexes were calculated (Table 3). The samples did not show significant differences among the indexes, except Pot_2 which refers to drinking water.

Table 3. Calculated Diversity indexes of sample.

	Capt_1_ Fev_20	Capt_2_ Fev_20	Guandu_ Jan_20	Capt_ Jan_20	Capt_1_ Mar_20	Capt_2_ Mar_20	Pot_1_ Mar_20	Pot_2_ Mar_20
Shannon	4.439	4.231	3.662	3.978	4.482	4.331	4.465	3.274
Simpson	0.958	0.946	0.923	0.948	0.959	0.953	0.958	0.825
Pielou	0.561	0.532	0.483	0.529	0.563	0.556	0.559	0.486

Even with the necessity to confirm the species, some disease-causing and pollution-indicator genera such as *Microcystis*, *Escherichia*, *Enterococcus* and *Prevotella* were found (Table 4). Increased some *Microcystis* species can induce hepatotoxicity [18]. Depending on which Escherichia *strain* is identified it can cause gastrointestinal ill-nesses, in addition to being an indicator of fecal contamination [19]. While some species related to *Enterococcus* genera can cause urinary tract infection [20], and some *Prevotella* species has been linked to several inflammatory diseases such as rheumatoid arthritis, periodontitis, and metabolic diseases [21]. Although we do not know exactly the species of the genera founded above, these results are worrisome and need exanimated carefully.

Table 4. Percentage of bacterial genera that can be pollution or disease indicators

	Capt_1_ Fev_20	Capt_2_ Fev_20	Guandu_ Jan_20	Capt_ Jan_20	Capt_1_ Mar_20	Capt_2_ Mar_20	Pot_1_ Mar_20	Pot_2_ Mar_20
Microcystis	23,46%	21,12%	11,64%	11,25%	29,04%	18,17%	30,73%	3,59%
Escherichia	2,09%	1,75%	21,49%	10,68%	4,13%	5,26%	2,46%	33,40%
Enterococcus	0,92%	0,69%	0,77%	0,53%	1,16%	2,50%	1,12%	9,67%
Prevotella	7,09%	6,30%	24,17%	16,91%	8,65%	9,13%	9,70%	9,05%

According to the cluster analysis, the Pot_2 sample was the one that stood out the most for belonging to an external group (Fig. 2), which was expected for being a potable water sample, while Pot_1 was grouped with a raw water sample (Chap_1_Mar_20). Furthermore, it was observed that the samples were grouped according to the date collection, which was also expected due to the similarity with the abiotic factors.

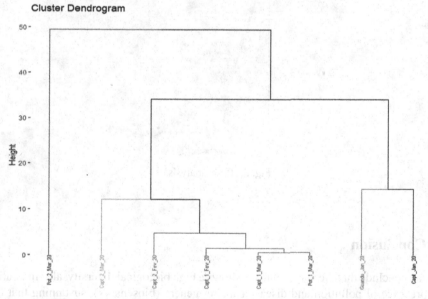

Fig. 2. Cluster analysis. In light blue, the Pot_2 sample stands out.

The PCA-Biplot (Fig. 3) shows three clusters and which variables differentiate the samples. We selected some genera that stood out in the PCA analysis for having greater contributions within the samples, such as: *Staphylococcus* - The genus includes commensals and pathogens of humans and animals. Some genera cause diverse infections in humans and have become increasingly antibiotic resistant over the past 70 years [22]; *Chthoniobacter* - is an important microorganism in the decomposition of organic carbon in the soil and exhibit significant positive correlations to various ARGs, which indicate that this genus is the main potential hosts for ARGs [23] and *Riemerella* - It is a genus that has pathological species normally found in birds [24]. Most of the selected genera have a high diversity of species that may or may not be pathogens, so we cannot determine the level of genera that is present in the samples.

The two first axes of the PCA analysis can explain 88.7% of the total variation in space. The first axis explained 65.1% of the total variation and the second axis explained 23.6%.

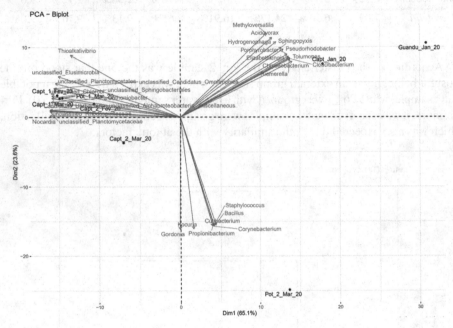

Fig. 3. PCA analysis.

4   Conclusion

We can conclude that the water analysis shows a high biological diversity, also indicating the presence of pollution and disease-causing genera (biosensors), sustaining that GR is a polluted river, however more studies are necessary to confirm the classification of the species. The taxonomic profile was similar between the samples, except for the Potable 2 point. This study demonstrates the potential of the methodology in monitoring and surveillance of aquatic environments. As perspective we will finish the recovery of genomes (MAGS analysis) from the samples to improve the taxonomy resolution and the results.

Acknowledgments. We thank CNPq, CAPES, and FAPERJ for their financial support.

References

1. CEDAE.COMPANHIA ESTADUAL DE ÁGUAS E ESGOTOS. Estações de Tratamento do Guandu e Laranjal. https://cedae.com.br/estacoes_tratamento#:~:text=Esta%C3%A7%C3%A3o%20de%20Tratamento%20do%20Guandu&text=A%20Esta%C3%A7%C3%A3o%20de%20Tratamento%20de,Queimados%20e%20Rio%20de%20Janeiro (2008). Accessed 20 Feb 2021
2. ANACIONAL DE ÁGUAS-ANA/MINIST. MEIO AMBIENTE. Plano Estratégico de Recursos Hídricos da Bacia Hidrográfica dos Rios do Guandu, da Guarda e Guandu-Mirim (2006)
3. IBGE. INSTITUTO BRASILEIRO DE GEOGRAFIA E ESTATÍSTICA. Rio de Janeiro. https://cidades.ibge.gov.br/brasil/rj/rio-de-janeiro/panorama. Accessed 25 May 2021
4. CEDAE.COMPANHIA ESTADUAL DE ÁGUAS E ESGOTOS. Qualidade da Água; Relatórios Guandu.2021. Available in: https://cedae.com.br/relatoriosguandu. Access in 02/20/2021.
5. Simon, C., Daniel, R.: Metagenomic analyses: past and future trends. Appl. Environ. Microbiol. **77**(4), 1153–1161 (2011)
6. Tringe, S.G., et al.: Comparative metagenomics of microbial communities. Science **308**(5721), 554–557 (2005)
7. Huson, D.H., et al.: Methods for comparative metagenomics. BMC Bioinform. **10**(1), 1–10 (2009)
8. Walter, J.M., et al.: Taxonomic and functional metagenomic signature of turfs in the Abrolhos reef system (Brazil). PLoS One **11**(8), e0161168 (2016)
9. Schmieder, R., Edwards, R.: Quality control and preprocessing of metagenomic datasets. Bioinformatics **27**(6), 863–864 (2011)
10. Coil, D., Jospin, G., Darling, A.E.: A5-miseq: an updated pipeline to assemble microbial genomes from Illumina MiSeq data. Bioinformatics **31**(4), 587–589 (2015)
11. Altschul, S.F., et al.: Gapped BLAST and PSI-BLAST: a new generation of protein database search programs. Nucleic Acids Res. **25**(17), 3389–3402 (1997)
12. Buchfink, B., Xie, C., Huson, D.H.: Fast and sensitive protein alignment using DIAMOND. Nat. Methods **12**(1), 59–60 (2015)
13. Huson, D.H., et al.: Integrative analysis of environmental sequences using MEGAN4. Genome Res. **21**(9), 1552–1560 (2011)
14. Scolforo, J.R., et al.: Diversidade, equabilidade e similaridade no domínio da caatinga. Inventário Florestal de Minas Gerais: Floresta Estacional Decidual-Florística, Estrutura,Similaridade, Distribuição Diamétrica e de Altura, Volumetria, Tendências de Crescimento e Manejo Florestal, 118–133 (2008)
15. Oksanen, J., et al.: Package 'vegan'. Community Ecol. Package, Version **2**(9), 1–295 (2013)
16. Glöckner, F.O., Fuchs, B.M., Rudolf, A.: Bacterioplankton compositions of lakes and oceans: a first comparison based on fluorescence in situ hybridization. Appl. Environ. Microbiol. **65**(8), 3721–3726 (1999)
17. Lindström, E.S., Kamst-Van Agterveld, M.P., Zwart, G.: Distribution of typical freshwater bacterial groups is associated with pH, temperature, and lake water retention time. Appl. Environ. Microbiol. **71**(12), 8201–8206 (2005)
18. Wei, J., et al.: Simultaneous Microcystis algicidal and microcystin synthesis inhibition by a red pigment prodigiosin. Environ. Poll. **256**, 113444 (2020)

19. Jang, J., et al.: Environmental Escherichia coli: ecology and public health implications—a review. J. Appl. Microbiol. **123**(3), 570–581 (2017)
20. Daniel, D.S., et al.: Public health risks of multiple-drug-resistant Enterococcus spp. in Southeast Asia. Appl. Environ. Microbiol. **81**(18), 6090–6097 (2015)
21. Iljazovic, A., et al.: Modulation of inflammatory responses by gastrointestinal Prevotella spp.–from associations to functional studies. Int. J. Med. Microbiol. 151472 (2021)
22. Moller, A.G., Lindsay, J.A., Read, T.D.: Determinants of phage host range in Staphylococcus species. Appl. Environ. Microbiol. **85**(11), e00209–19 (2019)
23. Jiang, C., et al.: Diverse and abundant antibiotic resistance genes in mangrove area and their relationship with bacterial communities-a study in Hainan Island, China. Environ. Poll. **276**, 116704 (2021)
24. Omaleki, L., et al.: Molecular and serological characterization of Riemerella isolates associated with poultry in Australia. Avian Pathol.**50**, 1–10 (2020)

Mesoscopic Evaluation of DNA Mismatches in PCR Primer-Target Hybridisation to Detect SARS-CoV-2 Variants of Concern

Pâmella Miranda[1,2]([✉]) [iD], Vivianne Basílio Barbosa[1], and Gerald Weber[1] [iD]

[1] Departamento de Física, Universidade Federal de Minas Gerais,
Belo Horizonte, MG, Brazil
{pamella-fisica,gweber}@ufmg.br
[2] Programa Interunidades de Pós-Graduação em Bioinformática, Universidade
Federal de Minas Gerais, Belo Horizonte, MG, Brazil

Abstract. Mismatches are any type of base-pairs other than AT and CG. They are an expected occurrence in PCR primer-target hybridisation and may interfere with the amplification and in some cases even prevent the detection of viruses and other types of target. Given the natural occurrence of mutations it is expected that the number of primer-target mismatches increases which may result in a larger number of false-negative PCR diagnostics. However, mismatches may equally improve the primer-target hybridisation since some types of mismatches may stabilize the helix. Only very recently have thermodynamic parameters become available that would allow the prediction of mismatch effects at buffer conditions similar to that of PCR. Here we collected primers from WHO recommendation and aligned them to the genomes of the current variants of concern (VOC): Alpha, Beta, Gamma and Delta variants. We calculated the hybridisation temperatures taking into account up to three consecutive mismatches with the new parameters. We assumed that hybridisation temperatures to mismatched alignments within a range of 5 °C of the non-mismatched temperature to still result in functional primers. In addition, we calculated strict and partial coverages for complete and mismatched alignments considering only single, double and triple consecutive mismatches. We found that if mismatches are taken into account, the coverage of WHO primers actually increase for VOCs and for the Delta variant it becomes 100%. This suggest that, at least for the moment, these primers should continue to be effective for the detection of VOCs.

Keywords: DNA mismatches · PCR primers · Mesoscopic models

Supported by organization Conselho Nacional de Desenvolvimento Científico e Tecnológico (CNPq) and Coordenação de Aperfeiçoamento de Pessoal de Nível Superior (Capes/Ação Emergencial, Brazil, Finance Code 001).

© Springer Nature Switzerland AG 2021
P. F. Stadler et al. (Eds.): BSB 2021, LNBI 13063, pp. 145–150, 2021.
https://doi.org/10.1007/978-3-030-91814-9_15

1   Introduction

The emergence of the pandemic of COVID-19 required the deployment of large-scale testing to control and monitor the disease. For this purpose, several protocols of PCR-based methods, mainly RT-PCR, were developed. Although RT-PCR is the gold standard molecular diagnostic, a few factors can interfere with its accuracy and performance such as sample quality and low amplification efficiency [9]. PCR efficiency in particular may be affected by destabilizing mismatches in primer-target. They may affect the ability of primers hybridise to the target, which may lead to non-amplification and, consequently, to non-detection. The influence caused in the hybridisation by mismatches depends on their length, sequential environment, position and number [4]. Even so, mismatches in primer-target duplex impact only the first few cycles of the PCR reaction [9]. They also affect the melting temperature, which is an important parameter to the primer design and is related to their stability and performance.

Here, we describe the evaluation of 21 primers and probes for RT-PCR recommended by WHO [1] in early 2020 to be applied to the detection of "original" SARS-CoV-2, which was evaluated in a previous work [5]. We collected those from Institut Pasteur, Department of Medical Sciences (Thailand) and National Institute of Infectious Diseases (Japan). We applied a mesoscopic model to calculate the hybridisation temperatures of alignments using a newly developed parameters for up to three consecutive mismatches [7]. The primers/probes were analysed regarding to SARS-CoV-2 variants of concern (VOC) classified so far: B.1.1.7 (Alpha), B.1.351 (Beta), P.1 (Gamma) and B.1.617.2 (Delta) variants.

2   Materials and Methods

Primer/Genome Sets. We collected 21 primers and probes from the summary of protocols recommended by WHO [1]. Regarding the genomes, we collected from GISAID [2] 7247 genomes of Alpha, 7497 of Beta and 2308 of Gamma variants in 7 April 2021, and 7943 genomes of Delta variant in 5 June 2021.

Primer/Genome Alignments. Primers and probes were aligned against each genome using Smith-Waterman algorithm [8], where AT and CG base pairs were given score 2, mismatches score -1, and no gaps were considered. Alignments were carried out regarding two strand configurations. The genome sequence as obtained from the database

$$5'-(\text{unmodified target genome sequence})-3'$$
$$3'-(\text{primer/probe sequence})-5'$$

and its complementary counterpart

$$5'-(\text{complementary target genome sequence})-3'$$
$$3'-(\text{primer/probe sequence})-5'$$

The alignments without mismatches were termed as strictly matched, those which contained up to three consecutive mismatches as partially matched and alignments with four or more consecutive mismatches were considered as not aligned. The limit of three consecutive mismatches is due to the available melting temperature parameters.

Calculating Hybridisation Temperatures. Hybridisation temperatures were calculated using

$$T_m = a_0 + a_1\tau, \tag{1}$$

where τ is a statistical index, which is calculated from the classical partition function of a model Hamiltonian, and a_0 and a_1 are regression coefficients obtained from a set of sequences containing up to three contiguous mismatched base pairs [7]. Moreover, the calculation of τ also generate the average displacement profile which shows the expected base-pair opening along the primer-target duplex. For a complete description of this calculation see Ref. [7].

Calculating Strict and Partial Coverages. We calculated the melting temperatures for the 21 primers/probes assuming a perfect hybridisation, which we called the reference temperature $T_{\text{ref.}}$, see Table 1. Alignments were carried out between primer and genomes of VOCs and kept only those with up to three contiguous mismatches. The coverage for strictly matched alignments C_{strict} was calculated as

$$C_{\text{strict}} = \frac{N_G - N_{\text{n.a.}} - N_{\text{MM}}}{N_G} \tag{2}$$

where N_G is the total number of genomes which are at least 25000 bp in size, $N_{\text{n.a.}}$ the number of genomes for which no alignment was found, and N_{MM} the number of genomes for which a partial alignment with up to three contiguous mismatches was found.

For partially matched alignments, we calculated the melting temperature T_{MM} taking into account the mismatches, and assumed the difference to the reference temperature T_{ref}.

$$\Delta T_{\text{MM}} = T_{\text{ref.}} - T_{\text{MM}} \tag{3}$$

Then, we calculated the partially coverage $C_{\text{part.}}$ as

$$C_{\text{part.}} = \frac{N_G - N_{\text{n.a.}} - N_{\text{low}}(\Delta T_{\text{lim.}})}{N_G} \tag{4}$$

where N_{low} is the number of primers satisfying

$$\Delta T_{\text{MM}} \leq \Delta T_{\text{lim.}} \tag{5}$$

Here, we use $\Delta T_{\text{lim.}} = 5\,^\circ\text{C}$, that is, we consider that primers with up to three consecutive mismatches with T_{MM} no more than 5 °C below the reference temperature $T_{\text{ref.}}$ are acceptable.

Availability. The software packages used for this work are freely available and can be found in https://bioinf.fisica.ufmg.br/software/analyse_mismatch_primers.tar.gz.

3 Results and Discussion

Mismatches in primer-template duplex may avoid the amplification and turn the PCR reaction non-functional [4]. However, in some cases, mismatches may contribute to stabilize the duplex, even the hybridisation may be greater considering mismatches in comparison to AT-rich primers with no mismatches [6]. Mismatches in the direction of 3′ end are more detrimental to PCR reaction [3], yet at and near 3′ end they may prevent false priming [6].

In Fig. 1a, a single AG mismatch is shown located at 3′ end, which yields a small surrouding perturbation (red line). Its temperature $T_{MM} = 55.3$ °C is

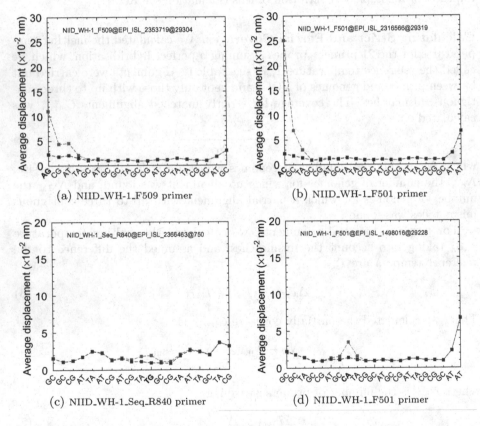

(a) NIID_WH-1_F509 primer

(b) NIID_WH-1_F501 primer

(c) NIID_WH-1_Seq_R840 primer

(d) NIID_WH-1_F501 primer

Fig. 1. Average displacement profiles. The blue line shows the displacement for full matched aligment and the red one for mismatched alignment. In each figure, the label shows the primer name, genome accession code and position, separated by the symbol @. (Color figure online)

out of the limit in relation to reference temperature $T_{ref.} = 63.6$ °C. In contrast, Fig. 1b shows a single AC mismatch at 3′ end, which yields a large end fraying and may impact in the DNA polymerase action, leading to a non-amplification. Nevertheless, its temperature $T_{MM} = 69.0$ °C is slightly lower than the reference temperature $T_{ref.} = 70.3$ °C, which indicates a feasible effective hybridisation. In Figs. 1c and 1d, we show single mismatches in the middle of the alignment, TG and CA pairs, respectively. Both single mismatches display a small perturbation to duplex in comparison to matched reference (blue line). However, TG sequence hybridises at a temperature of $T_{MM} = 45.4$ °C, considerable lower than its reference temperature $T_{ref.} = 60.2$ °C, whereas CA sequence hybridises at $T_{MM} = 71.0$ °C, which is slightly over to reference temperature $T_{ref.} = 70.3$ °C. The latter shows a feasible contribution of a single mismatch, which could stabilise the primer-target duplex without impact in the amplification.

In Table 1, we show both strict and partial coverages for the four variants of concern. In a considerable number of cases, the coverage increases considering mismatches and, in special for Delta variant, it increases to 100%.

Table 1. Results for 21 PCR primers and probes from WHO [1] recommendation. Shown are the reference temperatures $T_{ref.}$ and the range of strict and partially coverages for Alpha, Beta, Gamma and Delta variants of concern (VOC) genomes, respectively.

Primer/Probe	$T_{ref.}$ (°C)	Alpha variant		Beta variant		Gamma variant		Delta variant	
		C_{strict} (%)	$C_{part.}$ (%)	C_{strict} (%)	$C_{part.}$ (%)	C_{strict} (%)	$C_{part.}$ (%)	C_{strict} (%)	$C_{part.}$ (%)
NIID_WH-1_F24381	61.2	99.1	99.1	98.3	98.3	98.9	98.9	99.8	99.8
NIID_WH-1_F501	70.3	99.5	99.7	99.3	99.5	99.6	99.7	99.7	100
NIID_WH-1_F509	63.3	99.5	99.5	99.0	99.1	99.3	99.3	99.3	99.3
NIID_WH-1_R24873	61.5	99.7	99.7	99.3	99.3	99.4	99.4	100	100
NIID_WH-1_R854	61.7	99.1	99.1	98.3	98.3	99.6	99.6	99.4	99.4
NIID_WH-1_R913	69.2	99.3	99.8	98.1	98.6	99.7	99.9	99.8	100
NIID_WH-1_Seq_F24383	60.4	99.1	99.1	98.3	98.3	98.9	98.9	99.8	99.8
NIID_WH-1_Seq_F519	58.8	99.3	99.3	98.4	98.4	99.1	99.1	98.2	98.2
NIID_WH-1_Seq_R24865	60.1	99.7	99.7	99.3	99.3	99.3	99.3	100	100
NIID_WH-1_Seq_R840	60.2	98.9	98.9	98.1	98.1	99.6	99.6	99.5	99.5
WH-NICN-F	64.4	99.8	99.8	99.1	99.1	99.0	99.0	98.8	98.8
WH-NICN-P	51.3	99.9	99.9	98.6	98.6	99.2	99.2	99.9	99.9
WH-NICN-R	64.1	99.8	99.9	98.6	98.7	98.9	99.0	99.8	99.8
WuhanCoV-spk1-f	65.4	99.4	99.5	98.4	98.6	98.9	98.9	99.8	99.9
WuhanCoV-spk2-r	64.6	99.9	99.9	99.4	99.4	99.4	99.4	99.7	99.7
nCoV_IP2-12669Fw	54.3	99.8	99.8	99.3	99.3	99.2	99.2	100	100
nCoV_IP2-12696bProbe	67.0	99.7	99.8	99.4	99.4	99.9	99.9	98.9	100
nCoV_IP2-12759Rv	53.7	99.6	99.6	98.6	98.6	99.4	99.4	99.8	99.8
nCoV_IP4-14059Fw	54.8	99.9	99.9	99.5	99.5	100	100	100	100
nCoV_IP4-14084Probe	61.3	90.1	90.1	99.5	99.5	99.1	99.1	99.7	99.7
nCoV_IP4-14146Rv	54.8	99.5	99.5	99.5	99.5	99.8	99.8	98.0	98.0

4   Conclusion

We evaluated DNA mismatches in PCR-type primers/probes recommended by WHO to the detection of SARS-CoV-2 virus. We carried it out regarding the variants of concern classified so far. The impact caused by mismatches are not straightforward and a full evaluation can be carried out with a detailed calculation and up-to-date model parameters. Nevertheless, we showed that these primers are able to align to VOCs genomes in a high coverage and it is feasible a contribution of mismatches to primer-target hybridisation.

References

1. https://www.who.int/docs/default-source/coronaviruse/real-time-rt-pcr-assays-for-the-detection-of-sars-cov-2-institut-pasteur-paris.pdf?sfvrsn=3662fcb6_2
2. https://www.gisaid.org/hcov19-variants/
3. Bru, D., Martin-Laurent, F., Philippot, L.: Quantification of the detrimental effect of a single primer-template mismatch by real-time PCR using the 16s RRNA gene as an example. Appl. Environ. Microbiol. **74**(5), 1660–1663 (2008). https://doi.org/10.1128/AEM.02403-07
4. Elaswad, A., Fawzy, M.: Mutations in animal SARS-CoV-2 induce mismatches with the diagnostic PCR assays. Pathogens **10**(3) (2021). https://doi.org/10.3390/pathogens10030371
5. Miranda, P., Weber, G.: Thermodynamic evaluation of the impact of DNA mismatches in PCR-type SARS-CoV-2 primers and probes. Mol. Cell. Probes **56**, 101707 (2021). https://doi.org/10.1016/j.mcp.2021.101707
6. Mitsuhashi, M.: Technical report: Part 1. basic requirements for designing optimal oligonucleotide probe sequences. J. Clin. Lab. Anal. **10**(5), 277–284 (1996)
7. Oliveira, L.M., Long, A.S., Brown, T., Fox, K.R., Weber, G.: Melting temperature measurement and mesoscopic evaluation of single, double and triple DNA mismatches. Chem. Sci. **11**, 8273–8287 (2020). https://doi.org/10.1039/d0sc01700k, https://pubs.rsc.org/en/content/articlelanding/2020/SC/D0SC01700K
8. Smith, T.F., Waterman, M.S., et al.: Identification of common molecular subsequences. J. Mol. Biol. **147**(1), 195–197 (1981)
9. Suo, T., et al.: ddPCR: a more accurate tool for SARS-CoV-2 detection in low viral load specimens. Emerg. Microbes Infect. **9**(1), 1259–1268 (2020). https://doi.org/10.1080/22221751.2020.1772678, pMID: 32438868

Author Index

Printed in the United States
by Baker & Taylor Publisher Services